Alessandro Pirrone

Ecological Policy Handbook
Vol. III

Attractive solutions

INTRODUCTION

As we know, the perspective of the *Sustainable Scale* implies setting limits on the exploitation of natural resources and accepting our constrains. This approach is direct to solve the unprecedented global challenges we are facing, as well as future ecological challenges yet unforeseen. To better understand this point of view, we must introduce some characteristics, without the audacity to be exhaustive and definitive.

The major threat to a desirable human future is a level of material throughput which exceeds ecologically Sustainable Scale, as we have seen in the Ecological Policy Handbook - Vol. II Areas of Concern. So, a key characteristic of any enduring human society must be to ensure throughput remains within the regenerative and absorptive capacities of critical ecosystems while living off the income from natural capital and preserving the natural capital itself. This means *setting limits* on economic growth, on an expanding human population and on a wasteful technologies, to ensure these limits are respected.

The notion of limits is contrary to the currently dominant economic and cultural paradigm. Many of humanity's institutions, cultural myths and visual icons, assume and encourage continuous economic growth. Yet limits are a biophysical reality and will affect our well-being whether we recognise them or not. The mere mention of limits may conjure up fearful images of deprivation and return to a primitive existence. However, it is the continuous growth of absolute levels of material throughput that should be feared, as it will inevitably lead to deprivation and a reduced quality of life (see-Ecological Policy Handbook - Vol. I Overview and limits). This notion may also arouse concerns about the controls needed to maintain them. Centrally planned command and control economies did not fare well in the last century. Micromanaging economic activities would be costly, cumbersome, constraining, and counterproductive. Policy approaches are available that set macro-level goals and leave many of the allocation challenges to the market (we talk about that in Chapter III). The irony is that the longer we avoid limits, the quicker we will experience ever increasing levels of deprivation, thereby reducing our options for achieving Sustainable Scale. More authoritarian approaches will be needed if both our natural capital and options for self-management continue to dwindle. Accepting the goal of limited throughput can provide us with greater security and well-being than the illusory and impossible promises of continuous economic growth.

Ignoring the limit of material throughput that ecosystems can tolerate is a curious phenomenon, given our ready acceptance of limits in so many other spheres of life. We accept that we cannot fly like superman, be beamed star trek-style half way around the world or exercise our intelligence to solve any problem put before us. These are limits imposed by nature. We also observe stop signs when driving, refrain from murdering those we disagree with, and don't allow smoking in public places. These are social limits that enhance our safety and well being. We accept these limits for the common good and have learned we can enjoy ourselves and function freely within their confines. As with all limits that have value, they are generally "invisible". Yet the reality of biophysical limits to throughput is viewed as heresy. The purpose of respecting limits is to ensure our activities allow us to live off the income from natural capital, rather than drawing down the stock of natural capital itself.

We can do this in different way.

First of all, we should ensure that harvest of renewable resources, associated with specific throughput activities, rests below the natural regeneration rates of all critical ecosystem services. When a forest or marine ecosystem produces lumber or fish, their harvest should remain below the rate at which the lumber or fish are regenerated by these ecosystems. If the throughput or harvest exceeds the regeneration rate, we are exceeding Sustainable Scale and are decreasing the fund of natural capital upon which our harvest depends — the forest or fish, respectively. Continued liquidation of natural capital will eventually degrade the ecosystem's capacity to regenerate at the rate needed to maintain human needs, and result in deforestation or collapse of fisheries, as has happened around the world. This approach requires a lot of attention because we are reducing also the ecosystem resilience. Greater safety lies in erring on the side of preserving rather than degrading natural capital, especially when potential irrevocable changes in critical ecosystem services are at stake. Estimates of regeneration rates will be fragmented, incomplete and unreliable, dependent on a wide variety of difficult to foresee factors. As our scientific understanding of ecosystem services and dynamics improves, estimates may become more reliable. But some degree of uncertainty will always remain. Just how far below the regeneration rate the harvest should be, is a matter of optimal scale and is discussed below. It must also be considered that all services are critical, and the harvest of a specific resource (e.g., lumber) affects many other ecosystem services associated with the fund from which the stock is taken (e.g., the forest). Cutting trees not only provides lumber but also affects soil erosion, water retention

3

and a host of other ecosystem services provided by the forest. Those services that are important to human well-being, the loss of which would be irrevocable, are identified as critical. When applying this rule, consideration needs to be given to all affected critical ecosystem services, not just the resource of commercial interest. It is possible that the regeneration rates for some critical ecosystem services are exceeded, even though sustainable levels of resource extraction appear to be respected. Sustainability should be viewed in terms of all critical ecosystem services which are affected by extraction, not just the resource stock. For example, the harvest of lumber may be less than the regeneration rate of lumber, and by this criterion alone, appear sustainable. However, if the harvest involves multiple roads through the forest, then other critical ecosystem services such as flood control or habitat protection might be compromised. If they are degraded below the level required for human well-being, then the Sustainable Scale of the lumber harvest will have been exceeded. The emission of CFCs was considered safe because the compounds were inert and non-polluting. Yet, these very characteristics that made them appear safe at one level, were precisely those characteristics that made them dangerous in terms of ozone depletion in the upper atmosphere.

Second, we should keep our throughput rate so that emissions or wastes do not exceed the absorption or decomposition capacity of natural processes in a significant period of time. Here, the most serious examples of unSustainable Scale are the result of violating this rule. Wastes from industrial agriculture exceed Sustainable Scale for many local and regional sinks. Emission of ozone depleting compounds very rapidly resulted in destruction of the protective atmospheric ozone layer. Anthropogenic emission of greenhouse gases increased the concentration of these substances to levels unprecedented over the course of human civilisation, threatening the climate stability we depend on. There are two cases where the regeneration rate is zero which are scale relevant flags, calling for special considerations. The first case involves substances for which a critical ecosystem has zero absorptive capacity, and thus cannot be regenerated. Many synthetic organic chemicals fall in this category. In such cases there should be no production of wastes which cannot be broken down and absorbed. The second case involves non-renewable resources we describe after. When wastes cannot be broken down and absorbed, the need is to find substitutes which are biodegradable for the material in question. The only circumstance which might justify the continued use of a substance which cannot be broken down and absorbed, would be one where the benefits to basic human needs were immediate and overwhelming, and the ecosystem in question could be sacrificed, as determined by due process. Economic considerations should not constitute an exception to this point.

Third, when we have to do with any use of non-renewable resource, we should be coupled with investment in replacing it with renewable alternatives, or coupling the non-renewable usage with a renewable offset[1]. Examples might be using wood instead of plastics, or tree-planting for carbon sequestration, respectively. Our civilisation depends on a vast and diversified range of non-renewable resources and the challenge concerns the reduction and, if possible, the elimination of their use, in particular the most harmful ones. Many of these substances are toxic or damaging in use (e.g.,

[1] Daly, Herman (ed). Ecological Economics and the Ecology of Economics. Northampton: Edward Elgar, 1999

mercury, plutonium, fossil fuels), or their extraction destroys habitat and erodes biodiversity[2]. There is also a moral concern — if we use up valuable non-renewable substances, then we foreclose their use by future generations. We have an obligation to restrict use of valuable non-renewable resources and minimise their waste. Yet mainstream economics encourages rapid exploitation of these resources, relying on the substitutability of financial capital for any exhausted natural capital.

Dealing with non-renewable resources is not the current global economy approach. However, it's possible to program different strategies which would be helpful to contain this wide use of unsustainable resources, such as the ban of the most harmful compounds, research and develop renewable substitutes, recycling all the existing non-renewable substances so that raw materials are no longer needed, design goods which use nonrenewable resources for durability and which can be recycled. The only circumstance in which nonrenewable resource extraction might be justified would be if the benefits in terms of basic human needs were deemed sufficiently critical by due process; economic considerations should not determine exceptions to the rule. If continued use is deemed necessary, then serious research efforts to find substitutes should continue, and the extraction terminated as soon as possible.

Respect of justice and safety while ensuring ecological sustainability is another way to face the problem. The approaches described have a focus on ecological sustainability, and taken together would go a long way to achieve that goal. However, sustainability from the perspective of a global civilisation involves more than ecological

---

[2] Czech, Brian. Shoveling Fuel for a Runaway Train. Los Angeles: University of California Press, 2000

considerations. Social and ethical issues are also relevant; without respecting these concerns, ecological sustainability may not be feasible. The *Brundtland*[3] definition of sustainability reflects both perspectives in its emphasis on responsibility to future generations. Ethical concerns for future generations apply equally to current generations living in poverty. Meeting the full range of human needs for the global population through continued economic growth is no longer possible once limits are acknowledged. Sharing of resources derived from the income of natural capital alone becomes necessary to achieve social justice as well as ecological sustainability.

The concept of *optimal scale* integrates both criteria for sustainability, and identifies a number of additional issues to be addressed in implementing a scale perspective. We talked about the practical necessity of social justice to achieve and maintain ecological sustainability[4]. Optimal Scale is the key policy objective for a sustainable future. It integrates the goals of ecological sustainability and social justice, by focusing on the relative costs and benefits of marginal increases in throughput, where the costs and benefits include social and ethical as well as economic considerations. Optimal scale requires due process to ensure justice and safety, as well as ecological sustainability, are adequately considered.

Ecological sustainability is covered by the above three approaches, and some additional are needed to respect justice and safety concerns.

Justice requires *(i)* meeting the basic human needs of the current global population, including the poorest (in contrast to supplying

---

[3] https://sustainabledevelopment.un.org/content/documents/5839GSDR%202015_SD_concept_-definiton_rev.pdf

[4] Ecological Policy Handbook - Vol. I Overview and limits, see Dialogue with the Skeptics

luxuries for those already well off — the priority of the current economic paradigm); *(ii)* meeting the basic human needs of future generations (rather than the current practice of liquidating both non-renewable and renewable resources); *(iii) p*reserving biodiversity, thereby respecting the rights of non-human species, and recognising the delicate and intricate interdependencies of the web of life, of which humans are a part.

Safety Margin requires the identification of adequate: *(i)* review of the scientific knowledge available concerning the ecosystems affected by human activities and the criticality of the services they provide; *(ii)* an appreciation of the irrevocable nature of ecosystem change beyond a certain point, and the associated loss of life supporting ecosystem services; *(iii)* special precautions for making decisions about critical life support functions under conditions of uncertainty; *(iv)* identification of at least the minimum information required to make a safe decision; *(v)* ensuring adequate monitoring and enforcement mechanisms are in place to ensure that sustainability and social justice standards are respected; *(vi)* ensuring that an informed public participate in the determination of the above justice and safety issues. The description of these components of optimal scale is not an indication for any given human activity. The process of identifying optimal scale will be a complex political, social and scientific process, and undoubtedly entail many bumps and starts along the way. Given that many human activities are already unsustainable, the need to begin this process is urgent. Mainstream economic theory and practice places the highest priority on market allocation of resources. Distribution or justice concerns are secondary, and discussions of scale are almost nonexistent in mainstream discourse. Optimal scale requires a reversal of this priority ranking.

Ecological scale issues are most fundamental, relying as they do on laws of nature. Determining throughput levels that are within a Sustainable Scale range are therefore the first priority in planning a sustainable future. Scale *(i)* sets the limits; *(ii)* identifies the range of throughput that degrades critical life support services that are essential for human well being; *(iii)* identifies what is available to distribute; (iv) identifies what there is to allocate.

Just distribution requires ethical and social considerations of acceptable levels of inequity. Justice does not require equality of wealth, but fairness in distribution. However, if inequality is too great the discrepancy becomes unfair because it confers power and privilege that can be used by the wealthy to increase the inequality even more, and override the rights of others. Once sustainability criteria are set, ethical principles for just distribution needs to be determined before resources or wealth can be allocated.

Sustainability and justice goals set ecological and ethical limits, respectively. Once set, the issue of efficiently allocating the resources available can be determined, including the use of market mechanisms. Where appropriate, non-market policy instruments are also required (we will see later). Each of these major goals, *ecological sustainability*, *social justice* and *efficient allocation*, require separate and somewhat independent policy instruments to be achieved. The order in which these separate issues is addressed is important, as ecological limits must be set before issues of distribution can be addressed, and the limits established by both sustainability and justice concerns must precede the issue of allocation[5].

---

[5] Daly, Herman, and Farley, J. Ecological Economics: Principles and Application, Island Press, Washington, D.C., 2004

9

With this introduction, we hope to have identified the broad conditions necessary for ecological sustainability, as well as the social and justice conditions required. These requirements differ significantly from the priorities reflected in mainstream economics, public policies and many other areas of human activity. Changes are needed not only in the goal of continuous growth (to one of optimal scale), but also in the areas addressed below.

Chapter I
Vision for a sustainable future

The currently dominant economic paradigm contains a future vision of increasing material well-being, reliance on technological fixes to environmental degradation, and reduction of poverty and other social problems by continuous economic growth. Considerations of sustainable ecological scale indicate that this vision of the future is inconsistent with fundamental laws of science, ethical concerns for just distribution, and the evidence of what constitutes human happiness and well-being. Yet the notion of continuous economic expansion of the market economy as the solution to humanity's major challenges persists. If the current paradigm is not sustainable, then what is an alternative vision for humanity's future?

There are many possible visions of a sustainable future, which might differ in terms of the emphasis on technology, the level of material consumption, the size of the global population, the role of trade, the system of decision making, and in many other ways. Indeed, if people are to live in harmony with their immediate environment, different solutions will likely emerge based on the different biophysical characteristics of local and regional areas. Social and cultural diversity will both influence visions for, and be reinforced by, a sustainable planet.

Various groups have begun exploring these possibilities, but it is not the purpose here to choose among them, or to describe in detail what the most desirable vision might be. Such a process will inevitably be a slow evolution, integrating insights and aspirations from many people, and vary from region to region. So, we only try *(i)* to identify those characteristics of community and society that are essential for ecological sustainability, characteristics that, regardless of diversity and cultural preferences, must be respected, *(ii)* to suggest some approaches as to how these characteristics might be developed, *(iii)* to encourage the development of multiple visions of a sustainable future based on a Sustainable Scale perspective.

Quantitative growth involves an increase in size. Qualitative development involves an improvement in functioning. Over the last 250 years both quantitative growth and qualitative development occurred in some human societies as the result of economic growth. Greater comforts and well-being came in part from more productive farming, industrial production and trade. Quantitative development was so successful that it allowed global population to increase more than 500% in less than 200 years, and for this ever expanding human civilisation to impact every area of the planet. One of the greatest benefits of this growth in the human economy is the surpluses it provides for some, beyond the basic needs of food and shelter. These surpluses support innumerable qualitative developments — through improvements in the quality of life by removing the drudgery of earlier societies, increasing human health status and life span, and encouraging science and the arts to flourish — at least for some. The major benefits of economic growth has been in these qualitative improvements, not just in the increasing size of the economy itself. The surpluses from quantitative growth generate levels of material consumption beyond those required for human happiness and well-being. These surpluses, enjoyed by only a small portion of the world's population, are also generated in extremely wasteful ways, which unnecessarily degrade ecosystems. The maldistribution of material goods creates problems of excess

12

for some, and poverty for others. The relationship between growth and development is not a simple, linear one, as the current economic paradigm implies. Distinguishing between quantitative growth and qualitative development helps identify important differences.

Quantitative growth contributes to well-being only up to a certain level of consumption. Beyond this level additional quantitative growth does not contribute to qualitative development; it is at best wasteful, and at worst a major contributor to ecological degradation and social inequity

Qualitative development (not material affluence) is more relevant to human well-being once a certain level of comfortable sufficiency is provided by quantitative growth, and can occur with minimal throughput, thereby resulting in fewer threats to ecological sustainability than continued economic growth.

With this distinction, the broad *goals for a sustainable future* become clearer: *reduce quantitative growth of material throughput to remain within ecologically Sustainable Scale, and increase the qualitative developments which contribute to human happiness and well-being.*

One of the biggest challenges in making a shift to a sustainable future is in refocusing our policies and practices across a variety of fields, from economic growth to qualitative development. The focus on economic growth was appropriate for an earlier phase in human history when material goods were scarce, and more such goods contributed to human well-being. This is still true for meeting the needs of the world's poor. Continuing an exclusive focus on economic growth, especially in affluent countries, is counterproductive to a sustainable future. Balancing economic growth where it is needed (giving priority to meeting the needs of the poor, and in affluent countries focusing on maintaining basic infrastructure), with a focus

13

on qualitative improvements for all, is a more certain route to sustainability.

The last century and a half has been anomalous regarding the enormous growth in energy supply made possible by the use of fossil fuels. While fossil fuels will be available for several decades to come, they will soon become increasingly expensive, difficult to extract, and polluting. The required transition to renewable energy sources will be the first time in human history that an adjustment from one fuel source to another involves a decrease in energy intensity — the amount of work available from a quantity of fuel. Renewable energy sources have lower energy returns on energy invested than fossil fuels, so the implications of this physical reality are that either we prepare ourselves to invest considerably more financial resources to maintain the energy supply we currently experience, or we prepare ourselves to manage with considerably lower levels of total energy. It is highly unlikely that global energy supplies can continue to grow into the next century, as is currently assumed, without disastrous ecological impacts. Learning to live comfortably with less energy will become a necessity.

From a global perspective, optimal scale determines a sustainable level of throughput, considering acceptable risks, for a given population. The greater the global population, the lower the average per capita level of throughput at that risk level. If there is a fixed amount to go around, then it will have to be spread more thinly the more people there are to share with. Higher populations will thus push decisions about optimal scale either to lower levels of average per capita consumption, or to accept higher risks of exceeding Sustainable Scale if consumption levels are too high. The sensitive issue of global population policy will need to be part of the discussion regarding optimal scale.

The *Ecological Footprint Measure*[6] indicates that global throughput levels have been exceeding the annual income produced from natural capital for several decades. In 2019 the approximately 12.2 billion acres of productive Earth[7], divided by the 7.8 billion people who depend on it for their well being, results in an average of approximately 1.6 hectares per person as the *equal Earth share* available. Collectively we are currently using approximately 2.2 ha per person or over 20% more than is produced annually. These data indicate that global throughput currently exceeds Sustainable Scale as well as optimal scale by a considerable margin. Any decisions to reduce global population to achieve a higher level of per capita throughput would take decades to have an effect. Since the bio-productive land available is fixed (or actually declining as would be expected from the excessive drawn-down of natural capital reflected in ecological overshoot data), and the global population is going to rise by at least 2 − 3 billion over the next few decades, it appears the only way to reduce this ecological overshoot and approach ecological sustainability is to reduce our absolute level of throughput for the foreseeable future. While optimal scale is the ultimate goal for overall sustainability, the more immediate need is to move to an ecologically sustainable level of throughput. Population size is a sensitive cultural and moral issue for many people and nations. It is also difficult to predict with any accuracy more than a few years ahead. Whatever the size of future populations, it will have a dramatic impact on ecological sustainability and human well-being, and therefore deserves

---

[6] Ecological Policy Handbook - Vol. I Overview and Limits

[7] https://www.footprintnetwork.org/resources/glossary/

considerably more attention than it is currently getting. Any planning for a transition to Sustainable Scale must consider the impact of population on global throughput, the tradeoffs between sustainable levels of throughput and population size, and the considerable uncertainties associated with long term population projections.

There is little doubt that human activities are altering global ecosystems, and that this is happening more rapidly than previously understood. These changes are unanticipated consequences of economic growth, population increases and technological designs. Ecosystem science is in early stages of development. We do, however, know enough to realise that these global systems changes will have significant negative impacts on human well-being, although details regarding magnitude and timing remain uncertain. We know that multiple global systems are affected, that these systems are interdependent, and often respond in non-linear ways to the kinds of stresses now occurring, creating unwelcome surprises. We have unintentionally created a high risk, highly uncertain situation, with potentially disastrous and irrevocable consequences. Such challenges call for a cautious approach. International environmental treaties and many national and local policies have incorporated the *Precautionary Principle* as a way of dealing with these challenges but, unfortunately, most of them are only a printed version of good calls. It recognises our lack of knowledge regarding ecosystem dynamics, while appreciating our dependence on the enormous benefits they confer. It also recognises the potentially irreversible and extremely high risks involved in altering these systems as an unintended by-product of economic growth. Adopting a precautionary approach is an essential component of implementing a scale perspective.

Major threats to ecological sustainability are usually taken seriously only after costly disasters occur. Environmental problem solving has been end-of-pipe and symptom-focused, rather than proactive, preventive, and design focused. Now that the unintended negative consequences of our approach to economic growth are becoming clearer and the bases of our economic paradigms coming into question, waiting for disasters is neither necessary nor desirable.

The *focus on prevention* of scale problems should be a new economic and development activities, to be planned and implemented across a broad range of human activities.

Chapter II
Human happiness and well-being

Understanding the determinants of human happiness and well-being is important in the quest for ecological sustainability and social justice, because it helps us decide how to best use the limited material throughput available and identify what other non-material factors are important. Philosophers, theologians and social thinkers have wrestled with these questions for centuries. Their conclusions vary in detail but all agree that both material and non-material factors are important.

Literature Review[8]

Happiness has not been adopted previously, even in the reports issued by the United Nations. Happiness is one of the most controversial and popular topics of recent times. Recently many economists are looking for sources of happiness. Oswald AJ[9] pointed out that politicians made a mistake when they linked economic growth to happiness, where increased economic growth is not linked to individual happiness. And demonstrated through the results of which surveyed explained that industrial countries have not become happier over time, as well as the high rate of depression of such countries, especially the United Kingdom. Though the high income levels in the United States, which exceeds six times the United Kingdom, the suicide rate ratio is the same. According to Maddison[10] despite the improvement and increase of per capita gross domestic

---

[8] The Relationship between Happiness and Economic Development in KSA: Study of Jazan Region

[9] 2006

[10] 1991

product (GDP), happiness has not achieve any improvement through his study. While Easterlin[11] explained through a survey that increased income does not mean the existence of happiness during a certain period, where he presented his evidence after a series of studies nine different European countries, after the World War, that is, during the 1970s. He began his questionnaire in 1972 and ended it in 1991 and found the increase in per capita disposable income in this period was more than one-third. Also reported that happiness varies directly from person to person according to the income of the individual and upon the arrival of the community to a state of justice in the distribution of income; in this case will be achieved happiness. Happiness as a result of increased income and non-achievement will be caused by low income and lack of access to an appropriate level of well-being. While Hans Messinger[12] focused on gross domestic product (GDP) and pointed out that this was sufficient for the index of social well-being, as well as the measure of economic welfare, which some economists as Tobin and Nordhaus[13] explained the measures of economic welfare and defined the requirements of growth as any country in net national product.

During the beginning of the millennium Richard A. Easterlin[14] reached through his questionnaire that the data of the social survey of happiness according to social studies by psychologists such as marriage, divorce and disability most of the results are inaccurate because of their direct impact on happiness. Also in another study

---

[11] 1995

[12] 1997

[13] 1972

[14] 2002: 2003

Richard A. Easterlin[15] proved that there is a contradiction between happiness and income in the long term through his various questionnaires, which included many questions about happiness and satisfaction of life. Happiness from their point of view does not increase with the rising incomes of the country. The research continued through Easterlin using long time series to provide evidence that happiness and economic growth would not be realised in the short term but would extend in the long term and the relationship would be positive in developing countries. China has been excluded from these results because of the doubling of GDP growth rates, but the members of the community still do not feel happy and satisfied.[16] Differences of views that look at how to achieve happiness[17] need to achieve the health of the individual and all society; happiness lies in the welfare of society, and to compare happiness among countries through the success of their goals and trends, where the questionnaire of this study based on their own lives and social circumstances. The study showed contradictory results in the statement what is happiness?

Teng Guo, and Lingyi Hu[18] investigated the relationship between happiness and different economic variables in the United States. Their results showed that individual well-being can be predicted and measured. Authors concluded that there is an inverse relationship between happiness, unemployment and inflation, and this has been proven by many previous studies. Many researchers in develo-

---

[15] et al 2011

[16] Easterlin, 2015

[17] Richard Eckersley 2009

[18] 2011

pment science usually use the word happiness and relate it to personal well-being, which measures the satisfaction of their lives with happiness. Where Roy F. Baumeister[19] explained that there is a difference between happiness as life and happiness as meaning, by surveying a sample of 397 adults. He analysed various aspects of their lives with their behaviour, temperament, creative practice and more. The result of study showed that giving is the cause of happiness. Recently studies have found the role of technology, especially the use of smart devices in creating happiness to individuals.[20] Others[21] described the relationship between buying process and happiness whereas meeting needs considered as part of happiness, not only buying supplies but also highlighting the possibility of buying experience will contribute to increasing happiness.

On other hand Shoval D. H., Morag H.[22] analysed the possibility of a relationship between Jewish in Israel and Arabian students of nursing schools to find out whether there is happiness or not in dealing with them using questionnaire

So, some common findings emerge despite using different experimental and survey methods and exploring the issue in very different countries. One of the most common conclusions is that money or financial wealth is not the most important determinant. In this sense, knowing the meaning of *purchasing power parity*[23] could help us

---

[19] et al. 2013

[20] Rossi Kamal and Choong Hong (2015)

[21] Zining Peng and Maolin Ye (2015)

[22] 2017

[23] Purchasing power parity (PPP) is an economic theory that allows the comparison of the purchasing power of various world currencies to one another. It is a theoretical exchange rate that allows you to buy the same amount of goods and services in every country. https://www.thebalance.com/purchasing-power-parity-3305953

understand what are the determinants of happiness. Once we understand, it's suggested that an equivalent annual per capita income to $10,000 is enough to live in dignity, and beyond that, the happiness does not increase with the collection of material resources. But we need also a house, a car, we should pay for school, food, electricity and furnitures and so on. So maybe the world is not so *"inexpensive"* as we would like. However, the support of family, friends and community, a meaningful role in life, and basic freedoms are much more important at all levels of wealth beyond this range[24].

Economic theory endorsed the view that money makes you happier. Therefore, those with higher income are happier than those with less. Consequently, one can improve its life satisfaction by getting more money. Moreover, if public policy measures aim to increase the society income as a whole, then, there will be an increase in the well-being. In the same context, some analysis of developed countries, transition countries and less developed countries stated that there exist a positive association between happiness and income in the short run. According to the world happiness report, citizens of countries that are more stable economically have higher happiness levels. The economist devised a chart which presents that the link between per capita GDP and happiness is quite robust. Though, the importance of GDP growth in increasing living standards and creating new jobs, *Gross National Happiness* (GNH) approach stated that a country's progress does not only depend on the economic development but also on non-economical factors of well being. Many surveys deduced that contrary to what economic theory assumes about money and happiness is not correct. They found that

[24] Smil, Vaclav. Energy at the Crossroads: Global Perspectives and Uncertainties. Cambridge, MA: The MIT Press, 2003, p. 105.

more money doesn't make people happier. There are other life events such as marriage, divorce, disease, health … do have lasting effects on happiness. Therefore, knowing what influences happiness can help us all lead happier lives. However judgments of personal well-being differ from one person to another, each person's assessment of their own level of well-being is called subjective well-being. The raised questions are: what are the sources of happiness and how does happiness influence individual contribution in economic growth?

This relationship holds for both men and women, across age groups, and income levels, it has endured over the decades that researches have been conducted. It also supports the earlier observations of thinkers and philosophers, and is consistent with the intensely social character of human nature.

Another way of demonstrating this disconnect between material consumption and well-being is to consider the role of per capita energy consumption as it relates to various objective indicators. Energy use is a good index of general consumption as all consumption involves the use of energy. Infant mortality and female life-expectancy at birth are two of the best composite indices of well-being, integrating complex interactions among nutrition, health care and environmental exposure over an extended period.

The evidence from historical development, combined with future possibilities for improvements in structural and end-use energy efficiency and taken together with detailed studies from currently developing economies, suggest that approximately 100 GJ per capita per year primary energy is a reasonable benchmark for the average energy needs for people to experience a decent, modern quality of life. Our understanding and use of "a better life" is quite specific. It

23

refers to a world in which the basic material needs associated with housing, healthcare, adequate sanitation and effective transport are extended to everyone on the planet. It does not mean a TV in every room in the house, a new smartphone every year, three-car families or the "use once and throw away" practices that have become common in much of the rich world in the last 50 years. The question then becomes: how much energy is needed for a better life? A common measurement of energy is a gigajoule.[3] A single intercontinental long-haul flight from Cape Town to London requires an average of 40 gigajoules energy use per passenger. A physical labourer may deliver work that is roughly the equivalent of a gigajoule per year. If we take the United States, the current primary energy consumption is around 300 gigajoules per person per year — roughly similar to 300 physical labourers for every man, woman and child in the country. A more modest and energy-efficient economy, such as Japan or most European countries, averages around 150 gigajoules per person per year.[25]

Around this average benchmark, significant variation will almost certainly persist as a result of geography and legacy circumstances. But for many currently developed economies this conclusion highlights the potential for becoming much more efficient in resource use — and hence for reducing their demand overall — while for developing economies, it provides some sense of the scale of the energy system that needs to be put in place.

In a world of 10 billion people by the end of this century, aspirations for reasonably widespread prosperity will be associated with a global energy system approximately twice as large as at the begin-

[25] www.shell.com/energy-and-innovation/the-energy-future/scenarios/a-better-life-with-a-healthy-planet/_jcr_content/par/textimage_494361683.stream/1475857583070/830baf3cf119dcc4ee-c1e4a83cab7d243c18dfc7/scenarios-nze-brochure-local-print-awv9.pdf

ning of this century[26] (Japan and France used about 70 GJ/capita in the 1960s). Only marginal gains in these indicators occur if per capita energy consumption increases to 110 GJ (for comparison, the levels for France and Germany were in the 170 GJ/capita range in 2001). Beyond this level there are no further gains in well-being. Yet many affluent countries use in excess of 300 GJ per capita, while the world's poorest use less than 30. In addition, personal freedoms as measured by the *Freedom Index*[27] are compatible with societies using as little as 20 GJ/capita, demonstrating that basic freedoms do not require a high level of material consumption.

According to the world happiness report of 2017 which contained 155 countries all over the world, Saudi Arabia stands in the rank 37th internationally and 3rd among Arabic countries after Emirates and Qatar. This world happiness report is the fifth annual report since 2012. It has been based on six factors: per capita GDP, health, social support, trust and lack of government corruption, freedom and generosity.

Not only is a high level of material consumption unnecessary for happiness and well-being, but too much can actually be personally harmful. A strong orientation to materialism is associated with a variety of psychological and physical health problems[28]. Studies from many different nations, involving preschoolers to the elderly and both genders, show that placing a high value on financial wealth and material goods, regardless of income levels, is associated

---

[26] https://www.shell.com/energy-and-innovation/the-energy-future/scenarios/a-better-life-with-a-healthy-planet/_jcr_content/par/textimage_494361683.stream/1475857583070/830baf3cf119d-ce4eec1e4a83cab7d243c18dfc7/scenarios-nze-brochure-local-print-awv9.pdf

[27] https://www.cato.org/sites/cato.org/files/human-freedom-index-files/2019-human-freedom-in-dex-update-2.pdf

[28]Kasser, Tim. The High Price of Materialism. Cambridge, MA: The MIT Press, 2002

with higher levels of anxiety, depression, and low life satisfaction. Individuals with a strong materialistic orientation are more likely to be insecure, engage in antisocial behaviour, have personality disorders, and experience difficulties in intimate relationships. Materialistic values detract from personal happiness and well-being by reinforcing feelings of insecurity. Whatever positive feelings occur from material acquisitions are generally short lived, and require more acquisitions to reinstate the positive feeling. This creates an acquisition treadmill, characterised by unhappiness and insecurity which stimulates more acquisitions and subsequent insecurities. In the process, the kinds of interpersonal relationships that contribute to an enduring sense of well-being are neglected. Less empathy and intimacy are experienced, affecting others, including the children of those with high materialistic orientations. It is hypothesised that unmet security needs in childhood give rise to strong materialistic orientations, which are then passed on to the next generation.

And we arrive to the sources of happiness.

The *social factors* affect deeply people's well being and happiness. The factors that can be considered as sources of happiness and show the quality of life are: health, education, unemployment and marital status.

Health: since long decades many analysis presented the interaction between health and happiness. A good health will make people feel more cheerful and causes a greater contentment in their daily life. However having a bad physical or mental health leads to more restrictions in people's daily activities and causes a pessimistic view of life. Emotional health is also one of the strongest factors of happiness and well being. Therefore, developing the ability to cope with feelings of anxiety and depression is very beneficial. For example, a

26

serious accident or dangerous disease reduces one's happiness. However, when there is medication, health devices such as wheel chairs and a support network of friends and relatives, people can surpass this sadness and become happier. But even with adaptation, still there is a negative effect of poor health on happiness. In the United States and Europe, governments believe in the strong link between individual happiness and good quality of health care services received. People's satisfaction is playing an important role in improving the quality of health care reforms.

Education: happiness and education are strongly connected. A good education contributes significantly to personal and collective happiness. It is generally admitted that education improves people's lives in many aspects. Education enhances people's lives as higher educational attainment is linked to better career paths and is also believed to enhance outcomes in other life domains, such as health and relationships[29]. Education is considered as the most important activity in modern man life. In many countries it is one of the biggest items of public spending. The well-being of modern society is dependent not only on traditional capital and labour but also on the knowledge and ideas possessed and generated by individual workers. Education is the primary source of this human capital[30]. However, some recent empirical studies stated the opposite empirical relationship between education and happiness in developed countries. These studies observed that higher levels of education are allied with lower life satisfaction and subjective well being and this is

[29] Alfred M. Dockery. (2010). Education and happiness in the school-to-work transition. Centre foe Labour Market Research, Curtin University of Technology.

[30] Crocker, R. K. (2002). Learning Outcomes: A Critical Review of the State of the Field in Canada, Canadian Education Statistics Council, Ottawa

because of high expectations and aspirations of well educated people towards their lives' conditions and jobs. Therefore, we can assume that education affects well-being in part through its effect on income. Determining how does education influence happiness, depends on how one defines "education" and measures its personal happiness.

Unemployment: happiness and work was the subject of various empirical economic studies. Jobs are very important for sustaining individual's living, family and health which are the main elements shaping people's happiness. Accordingly, many studies stated a negative effect of unemployment on happiness. A high unemployment rate can be an indicator of a failing economic system and a call for policy action; this has been illustrated by the *Arabic Spring* where people suffered from difficult life conditions because of the high unemployment rate. In order to prevent high unemployment rates, governments should create new jobs. An important finding from the more recent literature is that there are large differences in the effect of unemployment among people—not all people are equally unhappy[31]. As some individual choose voluntary to be unemployed because they prefer to receive unemployment benefits and enjoy leisure time. Another important point to be mentioned is job security, which if declines it affects negatively workers happiness. High unemployment rates increase the fear of losing one's job and lead consequently to lower levels of happiness even for employed people. This means that employed and unemployed workers living in countries with high unemployment rates tend to be more unsatisfied. One of the most prosperous economies in the world today is Swi-

---

[31] Rainer, Winkelmann. (2014). Unemployment and happiness. Successful policies for helping the unemployed need to confront the adverse effects of unemployment on feelings of life satisfaction. IZA World of Labor 2014: 94

tzerland with a very low unemployment rate. The Swiss applies several tools in order to keep its economy strong, such as, implementing low interest rates, a short immigration policy and a strong banking system.

Marital status: generally, healthy and supportive relationships contribute greatly to higher happiness levels. The last world happiness report stated that, in western countries, having a partner has a strong positive impact on individual's happiness. However, marriage or divorce doesn't have the same impact on all individuals. If the person is satisfied with his or her life due to having a rich social network, marriage will not have a strong impact on his or her happiness. But, in the case of a lonely person, marriage will influence positively his or her happiness. Moreover, losing a partner of a happy marriage will impact negatively people's satisfaction. In a study of multiple regression analysis conducted[32] was found that marriage increases happiness equally among men and women and this positive impact is due to three facts[33]. First, marriage provides a financial satisfaction as married people combine two incomes and may enjoy a higher standard of living. Second, it leads to the improvement of health through the support encouraging partners to follow a medical treatment in case of illness, quitting bad behaviours such as drinking and smoking and helping spouse to follow a healthy diet. Finally, marriage provides greater emotional support which refers to being esteemed, cared about and valued as a person. However, the extent to which marital status is linked to individual's level of happiness is not the same for all nations. In Saudi Arabia where Sharia

[32] Steven Stack and J. Ross Eshleman (1998)

[33] Stack, S., & Eshleman, J. R. (1998). Marital status and happiness: A 17-nation study. Journal of Marriage and the Family; May 1998; 60,2; Research Library pg.57.

Law is applied, relations between men and women are strictly forbidden outside marital life. Unmarried couples are not allowed to live together or have intimate relationships outside the marital bond as is pregnancy. Traditionally, the ideal marriage was tribal, related families encourage their children to marry cousins or relatives in order to increase and strengthen the tribe. Another reason for such marriages was that families knew the background of the partner. According to Sharia law, a Muslim man can have four wives, provided that he can treat them equally. This practice is now decreasing mostly because women are becoming more independent and self-confident and many refuse to accept it. In Saudi Arabia, nowadays, both man and wife are increasingly going out to work, due to the elimination of restrictions on women working. Standing in the third rank of happiest Arabic countries, Saudi Arabia is one of the top ten Arabic countries in divorce percentage. This fact stresses on the question about the relationship between happiness and marital status in this country.

Infrastructure: is the backbone of many economic, social and political life activities in societies. Countries that do not care about infrastructure will be affected by the decline of its growth and development. China, Singapore, Malaysia and Thailand are considered the best performers among the world's economies today because they improved their infrastructure.

Freedom: it is always believed that people could live happier if their society practices freedom, equality and brotherhood. Freedom is considered as having the opportunity to choose and being able to. This means that being free requires the absence of restrictions in economic, political and personal life. Absence of economic and political restrictions can be used to measure differences across nations in their degree of freedom. According to previous studies, freedom is positively linked to happiness in rich countries. There are different opinions about the impact of freedom on happiness. Different philosophies express different effects and suggest different analysis. A conservative thought asserts that freedom might have negative consequences. According to this vision, people don't know what is best for them and their freedom has gone too far and about to destroy vital institutions. However, other schools have different opinion; they emphasise that only economic freedom which improves human well being but not political freedom. According to other thoughts, freedom leads to happiness under specific conditions such as people maturity. The comparative study[34] ranked nations according to the measure of two indexes *"opportunity to choose"* and *"capability to choose"*. These indexes were combined in a comprehensive measure of freedom. The rank order on that overall index shows that Nigeria, India, and China are the lowest scorers and the top ones are the US, Switzerland, and Canada[35]. These top countries are among the countries that have the happiest populations according to the world happiness report. A Comparative survey data[36] shows that the effect of political freedom is highly linked to an increase in happiness[37]. These studies were basically conducted among Western populations where political freedom is a constant background. In Saudi Arabia restrictions on dissent and freedom of

---

[34] conducted by Ruut Veenhoven (2000)

[35] Veenhoven, R. (2000). FREEDOM AND HAPPINESS A comparative study in 46 nations in the early 1990's. In Diener, E. & Suh, E.M. (Eds.), Culture and subjective wellbeing (pp. 257-288). MIT press, Cambridge, MA USA, 2000, ISBN 0 262 04182 0

[36] conducted by Inglehart et al.(2008)

[37] Bavetta, S., Patti, A. M., & Navara, P. (2014). Autonomy, Political Freedom, and happiness. Mimeo, PPE Research Unit, University of Pennsylvania.

31

expression are high. Though, the country combines a very low political freedom index with a fair degree of happiness. This fact can confirm the hypothesis that political freedom has a slight impact on people happiness.

Corruption: it is generally considered as an important factor which defines the quality of countries' governance. It doesn't have any geographical limits and exists everywhere. As shown[38], even though corruption is more spread in poor countries, it isn't restricted to specific region or levels of economic development[39]. According to economic theories, corruption at the macroeconomic level can be beneficial as well as harmful to economic progress and consequently can affect the *subjective well-being (*SWB). Boon vision asserts that corruption can serve to overcome institutional inefficiencies which in turn promote economic growth. Whereas, bone theorists propose that inefficiencies are originally caused by corruption itself and consequently restrict growth. In corrupt societies, wealthy residents have more freedom in their behaviour and decisions than poor ones because their wealth helps them in purchasing conveniences. Therefore, in these societies, income matters more than SWB. In general, corruption reduces institutional trust which in turn erodes SWB. Living in a country reigned by corruption and discriminatory institutions can lead to unhappiness. This is illustrated by a report released by the United Nations Human Development program about the case of Bosnia and Herzegovina in 2002. The reported survey showed that 70% of people in these countries strongly believe that their local authorities and international aid organizations are cor-

---

[38] by Abed and Gupta (2002),

[39] Abed, G., & Gupta, S. (2002). Governance, Corruption, and Economic Performance, Washington, DC: International Monetary Fund.

rupt. The report concluded that citizens of these countries are not happy because of corruption which in their point of view has broken all barriers and rules of life. As corruption represents a serious obstacle to good governance, economic growth and stability, international organizations such as United Nations and International Monetary Fund, launched guides of anti-corruption strategy in order to help countries enhance transparency and justice.

According to what has been mentioned above, social factors are the main source of happiness that drives economic development. Therefore, we can conclude that there is a strong relationship between social factors which are in our survey: health care, education, unemployment and marital status, and economic development. Governmental societies should accord happiness a major importance as an input not just an output. As happiness is now an accepted and crucial policy to implement and execute the objectives of the government targeting big aggregates such as economic growth or unemployment.

Mainstream economic theory is at variance with the disconnect between materialism and personal happiness. It assumes people are insatiably acquisitive, and that the greatest social good comes from individuals freely making market decisions in their personal best interests. Mainstream economic theory views people as highly individualistic, and the expression of this individualism in the market as the most certain and efficient way to achieving the greatest good for the greatest number. Mainstream economists have argued that considerations for others, or sharing of resources is actually counterproductive to market solutions to environmental degradation or social inequities. Anything that interferes with market operations — limits, regulations, quotas or notions of justice or altruism — are seen as distorting the market, and thereby reducing the wealth of the community. This view of a single dominant human need is overly simplistic in terms of the complex range and interdependence of needs identified by many researchers and thinkers[40]. It totally ignores the social component of satisfying human needs, and overlooks the personal and social benefits of ethical decisions and behaviours. It fails to recognise the non-material determinants of personal happiness, and the non-market factors which determine well-being. A great many ecosystem services upon which we depend for our survival are non-market services, and by their very nature always will be. In addition, many of the factors that determine our happiness and well-being are not part of the market, and never can be. The view of self-interest as the driver of the common good overlooks the benefits derived from a range of public goods, from a money system and sewers, to health care and education. It fails to recognise that in some circumstances individual preferences are in conflict with social goods[41] and that mechanisms other than the market are required to ensure personal happiness and well-being.

We are inherently social creatures and our sense of self-worth and happiness derives in part from comparing our self with others. If others' wealth or income is different from ours, then we are more or less satisfied with our own level by comparison, regardless of our absolute level (assuming a minimal amount to meet basic human needs). Consequently, if there are considerable differences in wealth

---

[40] Max-Neff, M. Development and Human Needs. In P. Enkins and M.Max-Neff, Real-life Economics: Understanding Wealth Creation. London: Routledge, 1992: pp. 197-213
Daly, Herman and J. Cobb. For the Common Good: Redirecting the Economy toward Community, the Environment and a Sustainable Future. Boston: The Beacon Press, 1989

[41] Daly, Herman and J. Cobb. For the Common Good: Redirecting the Economy toward Community, the Environment and a Sustainable Future. Boston: The Beacon Press, 1989

and income in a society, then there are also likely to be higher levels of dissatisfaction. Wealth and income are typically concentrated in a small proportion of the community, leaving the majority of people dissatisfied by comparison. Comparisons are generally made with those in a referent group, people we feel we should be equal with, or strive to emulate. Thus, even those who have considerable wealth and income compare themselves to others who are well off. Such comparisons are said to stimulate competition and raise the standards for everyone. One of the consequences of this process is that there is greater dissatisfaction for those who are left behind. Another consequence is that once the competition for more consumption starts, it triggers a continuing cycle. It is like standing up in a theatre to better see a performance. If everyone does it, then no one is better off, until someone stands on their seat and the cycle is repeated. Such increases in consumption to improve one's relative position add little if anything to one's enduring sense of well-being. But such competitive comparisons increase throughput and contribute to both ecological degradation and social inequities.

The relationships described above make it clear that a meaningful and satisfying quality of life is possible with considerably less throughput than now occurs in affluent and highly developed countries. A challenge for the future is to determine the best allocation of sustainable throughput levels to achieve the quality of life desired. What is clear is that leaving this allocation solely to market forces alone will not achieve Sustainable Scale!

The finding that non-material, quality of life factors are among the most important determinants of both subjective reports of human happiness and objective indices of well-being, is good news in terms of the opportunities to organise our societies from a Sustainable Scale perspective. Focusing on the dual objectives of reducing material throughput and increasing qualitative development should allow societies to remain within sustainable ecological scale while generating personal happiness and well-being.

The Sustainable Scale problems we face are unprecedented and serious, but there are solutions available and they are attractive for a number of reasons: *ecological sustainability and social justice* are more desirable than ecological degradation and social disintegration; while global limits on throughput are required, the *solutions are local* and will engage large numbers of people in a common cause; *everyone* will have meaningful contributions to make; the *policy solutions* available to solve Sustainable Scale problems have been used successfully with other problems; we have *experience* with most of them, and their familiarity should increase our confidence of success; we know how to produce goods *efficiently*, and meeting basic human needs with the necessary goods should be relatively easy; this can occur as well as increasing *leisure time*; individual freedom to pursue personal interests should increase dramatically with a focus on qualitative development within a *sustainable economy*; considerable *challenges* persist that will involve people's creative abilities and stimulate continuous qualitative development; the result should lead us to higher levels of personal *well-being* and stronger, more vibrant communities and *happiness.*

Our biggest challenges to implementing these attractive solutions is fostering the political will to put aside powerful vested interests who refuse to give up the short term benefits they derive from the current paradigm.

Chapter III
Public policies

This is the most inconvenient chapter to write, for a number of reasons that we believe it is useless to report. It must be added that in every part of the world, the human qualities of political decision-makers have been very rare, and only occasionally they performed for the best of people; these are those who are remembered by history, thus you can count them on the fingers of one hand. So, we just have to continue to suggest what is clear to many: what are the right public policies suitable to protect life from the suicidal dimension of the current globalist economic paradigm.

To be effective, policies must acknowledge that different goals are possible and that some are more desirable than others. These characteristics need emphasis because neither are part of mainstream economics, the dominant policy arena. Current economic theory and practice views growth as measured by increasing GDP as the ultimate and indisputable goal which will solve all environmental and social problems. Likewise, the market is placed in a hallowed position, raising it to the stature of a fundamental law.

"There is no alternative (to market capitalism)" - Margaret Thatcher, Prime Minister of UK, the home country of "de Rothschild family".

As mentioned, such a view is in sharp contrast with the fundamental laws of physics and the determinants of human well-being. Economic growth is only one of the contributors to well-being and only if it is limited by the laws of science. Markets are useful for efficiently allocating certain types of goods and services, and totally inadequate for allocating other goods and services that are equally important for human well-being. Economic growth and market dynamics, on their own, will only exacerbate ecological degradation and social inequities. Evidence reported in the Ecological Policy Handbook - Vol. I and Vol. II, and researches of all author across the world, suggest that global throughput currently exceeds both Sustainable Scale and optimal scale.

So, what policies will reverse the current processes causing this ecological overshoot and bring us to a sustainable future?

Before considering specific policy options, let us first review policy design principles that will encourage effective policy development. Six policy design principles to achieve optimal scale and social justice have been identified:[42]

1. Economic policy (which affects the level of material throughput) always has more than one goal, and each independent policy goal requires an independent policy instrument. The three goals proposed for economics are *sustainable ecological scale, just distribution, efficient allocation*. The absence of independent policy options in neoclassical economics for these distinct goals precludes the possibility of achieving ecological sustainability within the current neoclassical paradigm. Separate policy approaches are needed for each goal. How should these policy instruments be developed? This question is addressed by the remaining principles.

2. Policies should strive to attain the necessary degree of macro-control with the minimum sacrifice of micro-level freedom and variability. Sustainable limits of material throughput, or certain

---

[42] Daly, H.E. and Farley, J. Ecological Economics:Principles and Applications. Washington: Island Press, 2004

levels of just distribution, for example, require *macro-control* in as-suring that limits or standards are respected. This sort of macro-control should not be implemented like the several international and national environmental treaties and policies developed until "yesterday" because, as we know, they are bringing us to the end. These goals are important, not the particular means of achievement. Precisely how would be accomplished may vary from nation to nation and community to community. Centrally planned command and control policies are not needed. Markets can be useful in providing some level of micro-variability, but on their own cannot provide macro-control to ensure goals are attained (although this is precisely what neoclassical economic theory and practice assumes).

3. Policies should leave a margin of error when dealing with the biophysical environment. Ecosystem dynamics involve considerable uncertainties, and could involve irrevocable changes to increasing throughput demands. Adopting a precautionary approach would establish a safety margin between the demands we place on ecosystems and our best estimates of their capacities. Adoption of the *Precautionary Principle* is an essential step for ecological sustainability. This principle is included in several international and national environmental treaties and policies, but short-term economic considerations often overshadow its implementation (and, as we reported in previous Volume, it's also to considered the lack of an *efficient global power* or, let's say, the capacity to respect treaties by sovereign states after they voted).

4. Policies must recognise that we always start from historically given initial conditions. The world as it is today is our starting point and the basic institutions of governance and the market must be

gradually transformed to meet our policy goals. Gradualism respects what exists and what we have to work with, but it should not be an excuse for either inaction or diverting us from the goal of optimal scale.

5. Policies must be able to adapt to changed conditions. Conditions change, surprises occur, theories and policies are tested in the real world, and feedback provides new learning and insights. The process of *adaptive management* — changing policies as we learn more — should guide policy design and implementation for achieving ecological sustainability and social justice. In a world empty of material goods it made sense to increase what were scarce — material goods to improve human well-being. Success at accomplishing this now requires a change to dealing with what is becoming scarce because of that very success — natural capital and the ecosystem services they provide.

6. The domain of the policy making unit must be congruent with the domain of the causes and effects of the problem with which the policy deals. The idea is to deal with the problem at the smallest domain in which it can be solved. Problems should be addressed by institutions on the same scale as the problem. Garbage collection is a local problem and requires local policies. Climate stability and energy use are global problems and require global policy instruments.

We have a desperate need of a global perspective. Policies at the global level and the institutions to develop and implement them, are among the greatest organisational challenges we face in reaching optimal scale. Global environmental organizations such as UNEP, ECOSOC, Commission on Sustainable Development and environmental convention bodies are among the weakest of all international organizations. They generally require consensus rather than a majority to set policy, and enforcement mechanisms are often absent. Consensus is difficult because national interests often conflict, leading to compromises that fall short of the scientifically determined requirements. The fact that such international negotiations, led by diplomats, are generally held in secret, make it more difficult for civil society to monitor what is happening and have an influence on the proceedings. Fortunately there are a variety of policy options available that could be used to achieve optimal scale, many already in use in related areas. There are also many alternative means available to implement such policies in public and private sector activities. The magnitude and seriousness of Sustainable Scale problems have not yet been appreciated by governments or citizens, and without clear policy directions to target optimal scale as a priority, application of these options is spotty at best. A major hurdle that remains is acceptance of the reality that Sustainable Scale for human activities can only be achieved by acknowledging and respecting the biophysical limits of the ecosystems upon which we depend. Demonstrating that this can be accomplished in ways that satisfy the full range of human needs will hopefully help us overcome this obstacle.

What follows are some of the alternatives available to achieving optimal scale, identifying opportunities for contributions from various sectors.

Regulations

A regulation is a rule or order prescribed by an authority which controls or directs some activity, often in relation to a standard or target. Environmental awareness in the XX century led to a large number of regulations to protect people and the environment.

Bans, quotas and standards of various sorts have been ordered by governments, and fines or penalties are generally prescribed for violations. DDT was one of the earliest substances banned; individual paper factories have limits or quotas set for the amount of wastes they can discharge into a river; and emission standards have been prescribed for many industries. Other regulations require the use of prescribed technologies (e.g., best available control technologies, or BACTs, may be required to reduce pollution; the type of equipment used to harvest fish may be prescribed to limit habitat destruction). Regulations can address the depletion (input) or emission (output) end of the throughput flow. It is generally easier to control depletion activities, mainly because there are considerably fewer of them (fewer mines and oil wells than factories and automobiles). If input is limited directly, then output is also limited, indirectly. The quality of throughput must also be considered, as even small quantities of some substances can be highly toxic. In general, focusing on depletion rather than emissions is preferable. Another issue regulators must face is whether to focus on prices or quantity. If prices are used to limit throughput, for example by taxing them, then quantity of throughput can be affected, but only indirectly. In order to ensure that quantity remains within Sustainable Scale, constant adjustment in prices would be required as demand shifts with the impact of prices and other variables. If taxes are used to limit use, greater demand stimulated by scarcity and a growing population will lead to higher prices and higher consumption. Taxes would have to be raised to again limit quantity to the desired target. Such a system is not very efficient. As markets do not always respond in rational ways to prices, there would always be the risk of exceeding Sustainable Scale, making the result ineffective as well. And some market goods are relatively *inelastic*[43], and different taxation levels are unlikely to affect demand for them. Focusing on quantity in regulations provides greater certainty in controlling quantity within sustainable limits. Once those limits are set, prices can vary in the market, but the quantity of throughput will not be affected.

As ecosystems respond to quantity and not prices, the quantity approach to regulation is to be preferred, allowing market efficiencies to adjust prices.

Focusing on quantity rather than price is also consistent with the requirement that scale issues be determined prior to distribution or allocation issues. In addition, the quantity focus makes it clear that limits are real and important, and it avoids the illusion that if we are willing to pay high enough prices we can get as much as we want. Regulations can be quite effective at limiting pollution, and are helpful in managing renewable resources by respecting biophysical limits. But such command and control mechanisms are not always the most efficient ways of achieving the desired ends. Regulations have their own limitations as well. There are the issues of micro control, and property rights. In addition, once regulatory goals are achieved there may be no incentives to additional improvements. Regulations are generally command and control processes that involve micro control, violating the second rule of effective policy design, i.e., providing as much micro freedom as possible to achieve macro goals. Industries have often resisted command and control regulations, not only because the rules increase costs, but also because they interfere with how companies design their operations. Property rights present another challenge for regulations. Re-

---

[43] Inelastic is an economic term referring to the static quantity of a good or service when its price changes. Inelastic means that when the price goes up, consumers' buying habits stay about the same, and when the price goes down, consumers' buying habits also remain unchanged.

gulations which focus on depletion rather than emissions are generally preferable as outlined above. Regulating depletion involves focusing on sources rather than sinks. However, resources are generally owned, whereas sinks are not. Regulations which limit depletion therefore can be viewed as interfering with property rights, and are often resisted on such grounds. One way of reconciling the attractiveness of focusing on depletion and recognising property rights is to acknowledge that property rights are a "bundle of rights." What the resource owner is asked to do is give up one stick from the bundle — the right to determine the rate of extraction of the owned resource. The owner still receives payment for the resource and otherwise enjoys the rights of ownership. Property rights are social institutions that entitle one individual to a right, and create obligations for others to respect the right. The third party involved is an authority (the state) to ensure that the obligations are fulfilled. There are three types of property rights that are useful to keep in mind when considering how policies might be designed to ensure a scale perspective: *(i)* a *property rule* exists if one person is free to interfere with another, or free to prevent interference. Putting up a "no trespassing" sign indicates the right to prevent access, *(ii)* *liability rule* exists if one person is free to interfere with another, or to prevent interference, but must pay compensation for interference. An example is when a state appropriates land from an owner, but pays compensation; *(iii)* an *inalienability rule* exists if a person is entitled to either the presence or absence of something, then no one is allowed to take it away for any reason. The right to a safe and healthy environment might be such a right.

These rights are independent and may act in combination. It is also helpful to keep in mind that property rights need not be private property rights. Properties may also be collectively owned or owned by the state. Recent international treaties such as the Montreal Protocol and the Kyoto Accord recognize the need for property rights owned by the global community. Another potential limitation of regulations is that they may be designed so as to discourage any improvements beyond the limits or standards set, even if more improvements are ecologically desirable and technically feasible.

Then, there are other examples of some policy instruments available to assist in the implementation of a scale perspective.

Bans

A ban is a regulation that removes a substance from circulation, thereby eliminating throughput of a particular type. A ban is the simplest solution to establishing Sustainable Scale when the absorptive capacity of an ecosystem for a particular substance is zero. If an emitted substance cannot be absorbed or broken down through natural process, it accumulates in the environment where it causes damage. DDT, leaded gasoline, and CFCs were all found to cause damage to critical ecosystems, and all have been banned in many developed countries. Some of the earliest bans can be traced back more than 2500 years, when hunting certain animals was banned in India. Bans can take many forms: they can be total or partial; they can focus on production or consumption; they can be temporary or permanent; they can be graduated in time or magnitude; they can be supported by incentives or penalties. Many substances have been banned, and bans are used in a wide variety of situations — from local, seasonal bans on hunting and fishing, to global treaties imposing bans on specific compounds or activities. Bans can be very effective policy instruments. The Montreal Protocol has resulted in a dramatic reduction in the production and use of various ozone depleting compounds. DDT and leaded gasoline have largely been removed from use in many developed countries. They are a potentially powerful solution to serious ecological problems, and allow for considerable flexibility in design. The ultimate test of success is if they reduce throughput to a safe level below that required for regeneration of absorptive capacities of critical ecosystem services, including the micro-environments within biological organisms necessary for health. Bans also have the advantage of a long history at the local and regional levels, so policy makers can make use of previous experiences to apply them in new situations. They are easier to implement if there are healthy alternatives to the banned substances. Banning CFCs was resisted by producers until an alternative compound was developed. Another obstacle to bans is the immediate financial costs to users. Leaded gasoline, for example, is more expensive than unleaded, so it continues to be used, especially in poor countries although the harmful health effects are well known. Bans also have limitations to their effectiveness. A general limitation is that they are usually applied only after catastrophic environmental damage or loss has occurred. This can be corrected with determination and some foresight if we can learn to apply bans preventively. The resistance to bans stems from parties with commercial interests from both the production and consumption side, and may include such diverse groups as poor farmers avoiding malaria, residents of wealthy countries who often unknowingly rely on many contaminated products, and multinational corporations who produce the substance. Even with successful applications there have been difficulties achieving zero throughput due to cheating and technical and political slippage. Despite the general success of the Montreal Protocol, black market compounds are still traded, and there are requests for exemptions for some compounds. If bans cannot achieve zero throughput, Sustainable Scale will eventually be exceeded, regardless of the level of reduction achieved; exceeding Sustainable Scale may be significantly delayed but not avoided. When no alternatives are readily available, or if they are available but costly, then it is difficult to obtain agreement from the many nations involved, many of whom have competing interests in the matter. Lack of complete scientific understanding about the impact of certain substances can also be an obstacle to implementing a ban. Many substances that are known carcinogens or enzyme disruptors, for example, which might be considered for banning, have little or no research examining their safety. Tens of thousands of such substances are currently in use, and it is not feasible to conduct definitive research for all of them. The long term impact of these substances on human and ecosystem health may take decades to understand. Some of the more dangerous of these substances might be considered for banning, applying the precautionary principle. At the very least, a Sustainable Scale perspective could be used to consider bans of new substances prior to their being introduced, especially those which are used in frivolous ways. Bans are essential to maintain Sustainable Scale when emissions cannot be absorbed or broken down. Research and design approaches that seek alternatives for existing substances for which critical ecosystems have zero absorptive capacity are urgently needed. If alternatives are not available, changes in market and social values may be required to accept doing without the non-essential but dangerous substances. The greatest challenge from a Sustainable Scale perspective may be learning to accept preemptive bans — banning certain classes of substances from being introduced, unless it can be reasonably de-

47

monstrated that the substances are biodegradable or can be totally recycled. Strengthening the enforcement mechanisms and penalties for banned substances is also important, especially when critical global ecosystems are affected. If ecosystem capacity to absorb certain substances is at or near zero, then any leakages into critical ecosystems are unacceptable.

Quotas

Quotas are partial bans. They involve regulatory limitation of the absolute amount of a substance into the human economy; they are a strategy to establish the maximum allowable throughput of a substance, and could be very effective in ensuring specific substances only enter the economy at a sustainable level. Quotas may be preferable to a complete ban if there is evidence that some levels of throughput can be safely absorbed by the ecosystems they affect. This safe level of throughput allows the benefits of the substance to be made available. Quotas should not be used unless there is adequate proof that safe levels are indeed possible, and often there are disputes about this issue. The two categories of quotas are depletion and pollution or emission quotas.

A *depletion quota* recognises there is some value in having the substance in the human economy, but that there are also associated problems with having too much. Sometimes, depletion quotas are imposed strictly for economic reasons, as when OPEC puts a quota on oil depletion. This is very different from imposing a depletion quota for ecological reasons, whereby the quota or limit is determined by the impact of throughput levels on ecosystem functioning. Depletion quotas may be set for renewable or non-renewable resources. Imposing limits on resources at the point of depletion or extraction is generally more efficient than imposing limits on emissions, as there are generally more sources of emissions. Limiting depletion indirectly limits emissions. However, there are also many ecological, as well as social justice, issues involved with depletion of various natural resources. Mining and logging operations can disrupt habitat for many species, contribute to soil erosion and flooding, and displace people living on the land. The depletion process itself is a sustainability issue, irrespective of any emissions from the materials depleted in terms of their impact on ecosystems. Standards regarding depletion processes may be as important as depletion quotas in terms of ecosystem functions (e.g., there are now standards regarding the use of certain techniques for harvesting fish quota set). Depletion quotas for renewable resources such as fish and game are some of the oldest regulatory actions. Depletion quotas allow the species to reproduce at an adequate level to replace the stock taken. This maintains a somewhat constant stock of the renewable resource without exhausting it, and in theory maintains an optimal scale. Depletion quotas are used on a variety of renewable resources such as wild animals, fish, timber, water, soil, etc. Depletion quotas for renewable resources are increasingly important as the human appropriation of nature continues to expand. Determining how a host of renewable resources can in fact remain renewable and not be exhausted by human use, is a daunting task. To date we have not had notable successes with setting quotas for various fisheries, which collapsed despite the quotas established. These experiences point out the difficulties we have in understanding various ecosystem dynamics and the importance of applying the precautionary principle when setting quotas in conditions of uncertainty. From a Sustainable Scale perspective there should be no depletion of non-renewable resources. However, reliance on many of these resources is widespread, and depletion quotas could be a useful approach to eventually reaching the goal of zero depletion or throughput. Such an approach would also require that some of the profits from depletion were used to find a suitable substitute. The use of carbohydrates to replace petroleum in the manufacture of certain plastics is one example of the possibilities. From a Sustainable Scale perspective, the transition to renewable substitutes should occur as quickly as technically possible. At the very least, it should occur before the known reserves of the non-renewable resource are totally depleted. This would ensure, in theory, that an equivalent resource would be available for people in the future.

Quotas may focus on *pollution or emissions* as well as depletion. An emission quota sets a limit on the amount of emissions of a particular substance that is permitted. Pollution may be a problem generated by the use of naturally occurring substances, such as fossil fuels, or man-made substances, such as DDT. The Kyoto Protocol uses emission quotas to limit the amount of greenhouse gases from a variety of sources. Quotas are also important policy tools for dealing with manufactured substances. Modern chemistry has developed thousands of substances which nature has never been exposed to, and for which it has not evolved methods of adaptation. Many of these substances are toxic, such as plutonium and DDT, or quickly degrade ecosystems once they enter the economy. The size of the human population is a major factor in the overall scale of the human economy (*see Population and the IPAT Equation in* Ecological Policy Handbook - Vol. II and Vol. I, respectively). One approach to reduce the scale of the human economy is to reduce the size of the human population. A novel approach to encouraging a decline in population toward some (to be determined) global target is the concept of human birth quotas[44]. Once a total target was set, the total number of births desired could be determined and quotas issued to every woman. These quotas could then be traded or sold if different women wished more or fewer children than their quota allotment allowed. To create a sustainable world, Herman Daly has proposed the idea of a *Steady State Economy*[45], one which does not focus on continuous economic growth but on qualitative development, once a Sustainable Scale is reached. To achieve such a state would require the following quotas: *(i)* constant stock of humans, *(ii)* constant stock of goods for providing services and meeting need, *(iii)* sufficient and sustainable levels of the above, *(iv)* throughput reduced to the lowest level to maintain the above. Obviously there are tradeoffs among these quotas; with a limited and sustainable level of throughput (stock of goods), the level of "sufficiency" will vary depending on the size of the population sharing such goods. The concept of optimal scale as a key policy priority recognises the interplay between these quotas, and also the importance of defining these quotas through a consensus process. Quotas will play a key policy role in achieving Sustainable Scale.

Quotas have proven essential to limiting environmental damages from potential overuse of renewable resources, as well from man-made substances such as DDT. Quotas could be used as policy instruments to set limits on the use of certain non-renewable resources as the first step to eventual elimination. Quotas can be particularly effective where the jurisdiction in which they are applied have and use strong enforcement powers. When moving from a total ban

---

[44] Boulding, K. E. "The Economics of the Coming Spaceship Earth." Pp. 3-14 in H. Jarrett (ed.), Environmental Quality in a Growing Economy. Baltimore: John Hopkins University Press, 1966

[45] Daly, H.E. Steady-State Economics, (2nd edition). Washington: Island Press, 1991

51

to a partial ban or quota, there is always the issue of how much is enough and the competing views on this from different perspectives. A potential problem with quotas is that the scientific data may not be available to accurately identify sustainable throughput levels. There are many examples of inadequate quotas allowing continued ecosystem degradation, or even collapse, as with various fisheries. Where there is uncertainty, political decisions often favour the higher levels of throughput (contributing to more economic growth) rather than applying the precautionary principle and starting with lower levels of throughput (lower quotas). Such an approach could gather more data regarding ecosystem impact and adjust the quotas in the future. All quota mechanisms would be strengthened by including such a monitoring and adaptive management approach. Setting quotas in one jurisdiction does not affect emissions from elsewhere. Transboundary pollution is an increasing feature of global environmental problems, and the challenge of establishing quotas is made more complex. For problems of a global nature, complex international negotiations can take years, delaying the implementation of agreed quotas.

Because of the political nature of these international agreements regarding quotas, the quota levels can be based on political rather than scientific considerations. Quotas can be set which exceed Sustainable Scale. This has occurred with certain fisheries, and in the case of greenhouse gas emissions. When critical ecosystem functions are at stake the precautionary principle is available to compensate for scientific uncertainty. Too often it looses out to political expediency and economic interests.

Standards

Standards are prescribed levels of performance enforced by law. A wide rang of such standards were enacted in the latter part of the XX century as a response to growing awareness and concern over environmental pollution. Various national environmental protection agencies were established around the world from the 1970s on, and implemented a wide variety of environmental standards to control pollution. *Ambient standards* regulate the amount of pollutant present in the surrounding environment (ambient) such as parts per million (ppm) of dissolved oxygen in a river, sulphur dioxide (SO_2) in an air shed, or ground level ozone levels. Measures are often an average (e.g., over a 24 hour period or per year), as concentrations vary by time of day and by season (e.g., due to weather changes). The level itself cannot be directly enforced, therefore the sources of the pollution must be found and regulated to be sure that the ambient standard is met. The US *Clean Air Act*[46], for example, sets ambient standards for six criteria pollutants in a region. If a region is in violation, they must come up with a plan to attain compliance. *Emissions standards* regulate the level of emissions allowed such as emissions rates (pounds of SO_2 per hour), concentration (ppm of biochemical oxygen demand - BOD - in wastewater), total quantity of a pollutant, residuals per unit of output (SO_2 per kWh of electricity), residual content per unit of output (sulphur content of coal), or percentage removal of pollutant (90% of SO_2 scrubbed). Emissions standards do not guarantee a specific ambient level of pollution. Weather conditions affect the concentrations and human behaviour affects pollution levels. *Technology standards* require polluters to use certain technologies, practices or techniques. Whereas emissions stan-

---

[46] https://www.epa.gov/clean-air-act-overview

dards require polluters to meet a goal for the level of pollution, but give the polluter freedom to choose the technology used, technology standards require a specific technology. For example until 1990, electric utilities were required to install scrubbers with 90% efficiency ratings. The US requires catalytic converters in autos. The 1972 *Water Pollution Control Act Amendments* set a goal of zero discharges by 1985, and used technology based effluent standards (TBES) — a combination of a ban and a standard. The EPA determines the "best practicable technology" and sets standards assuming that firms are using that standard. Often, as in the *Clean Air Act*, the government mandates that the *Best Available Control Technology* (BACT) be used[47]. Banning certain technologies is another way of establishing a standard. Clear cut logging has been banned in certain jurisdictions and long line drift nets have been banned for certain fisheries. The generation of electricity with nuclear fission has been banned in some European countries.

Like the many other policy tools, standards can be very effective in reducing pollution of various types; they are often used in conjunction with other policy instruments such as bans or quotas. There are many flexible approaches to standards and considerable experience has accrued with regard to their use. One of the potentially negative aspects of standards is that they have often been of a command and control nature; that is, they prescribe not only, or even necessarily, a goal, but a specific means of achieving that goal. This "one size fits all" approach is not always the most effective or cost-efficient. Enormous amounts of financial resources have been expended by business and industry to comply with environmental standards by retrofitting existing infrastructures. In addition to resi-

---

[47] https://www.epa.gov/nsr/best-available-control-technology-bact-applicability

sting the imposed costs these standards require, business and industry have also objected to being told precisely how to achieve the desired goals. If standards can be set in terms of clear, measurable goals, business and industry prefer to have the flexibility of working out the methods for achieving those goals. Another problem with the command and control standards is that once achieved there is no incentive for exceeding the standard and providing even greater environmental protection even when this is possible. Incentives to exceed standards can be used to this end. Standards have been used successfully with a range of local and regional environmental problems. However, the level at which standards are set can have dramatic impacts on other levels. For example, setting standards at the national level for vehicle fuel efficiency can lead to increased vehicle use, exacerbating the problems at the regional and global levels through increased levels of throughput. As with any policy instrument, the key criterion from a Sustainable Scale perspective is whether the policy results in maintaining throughput within the sustainable range. To date standards have not been used to tackle global problems of Sustainable Scale. It remains to be seen if they can be successfully applied in these areas.

Emissions Trading Permits

Emission trading permits are regimes in which governments issue or sell permits to allow certain levels of emissions of potentially harmful substances, where the permits can be sold or traded amongst the parties with the permits. The theory is that emissions within such a regime will be kept within a desirable limit. Emission trading systems involve establishing quotas within which the trades may occur; consequently, the strengths and limitations which apply to quotas also apply to emission trading systems. In certain applications the emission limit is set in terms of absolute amounts, in other applications the emissions are set in relative terms. Thus specific applications can have dramatically different results in terms of their effectiveness in achieving Sustainable Scale.

In a *cap and trade system*, the total emissions are set or capped, and the total permits equal to this cap are issued among firms generating the pollutant. Firms can then trade these permits among themselves: those firms generating more emissions than their initial allocation would buy permits from firms that generated fewer. The total number of permits remains stable, but the distribution of permits among firms would change with the market. Number of permits can also be reduced from year to year to accomplish a reduction of emissions. This approach was pioneered in the United States to deal with the problems of acid rain. It has been used most successfully for sulphur—dioxide and nitrogen—oxides from power plants in the US and Denmark, where the quota was set at a declining amount over time to allow investment in technological change. For both these programs the emissions were consistently below the quota. In some programs non-compliance results in heavy fines which are more costly than purchasing permits. Cap and trade systems have been applied to a variety of pollutants, including sulphur dioxide, nitrogen oxide, lead in gasoline, ozone depleting compounds, oxides of nitrogen and carbon. Such schemes have been used in a variety of countries as well. These systems have generally been both effective and cost-efficient. Significant reductions in emissions have been achieved, and generally at much lower costs than originally anticipated. Today, cap and trade is used or being developed in all parts of the world. For example, European countries have operated a cap-and-trade program since 2005. Several Chinese cities and provinces have had carbon caps since 2013, and the government is working toward a national program. The Kyoto *Protocol* has endorsed the use of a cap and trade approach for carbon emissions to address climate change.

Emissions trading systems have proven effective methods for reducing a number of pollutants in a variety of countries. Cap and trade systems are of greatest interest from a Sustainable Scale perspective, as they involve setting an absolute limit, or cap, to the level of emissions permitted. This widespread experience and success with cap and trade systems is encouraging in terms of their potential applications in dealing with issues of Sustainable Scale. Given the flexibility they allow and their potential for combination with other policy instruments such as fines and standards, they hold much promise for the future. Emission trading schemes that do not set absolute limits to the amount of pollutants permitted will not contribute to Sustainable Scale. In fact, increased efficiencies from emission trading schemes without absolute limits, run the risk of actually leading to increased levels of emissions[48]. Furthermore, emission trading quotas have another limitation in common: the guarantee of staying within the Sustainable Scale. We know cap and trade systems are effective and efficient tools to significantly reduce pollutants affecting ecosystem functions. But unless the target levels of pollutant throughput are within Sustainable Scale, that goal will not be achieved. The emission targets for the initial Kyoto *Protocol* are significantly below that required to stop or reverse climate change, but at least it is a start. Emission trading schemes provide a legal right or permission to pollute, to use the global commons as a dumping ground for waste. This is not a problem from the per-

---

[48] Jevons Paradox

spective of ecological sustainability as long as the emissions are below the rate at which the sink can be renewed. There is also an issue of justice involved in terms of how this right is distributed. Governments may sell permits on behalf of their citizens, or they can grant permits outright on a *first come first* serve basis, or on the basis of *previous emission histories*. Businesses and industries, as well as developed countries, generally prefer the latter approach as it allows the largest historical emitters to receive the most permits. Recognising every person's right to clean air would lead to an allocation of emission permits to different countries based on population. This approach is preferred by countries with large populations, many of which are among the less developed nations of the world. Under this scheme, emission permits would go to countries based on population, industrialised countries or countries with large transportation infrastructures, would have to purchase emission permits from those with large (and generally much poorer) populations. The refusal of the developed countries to accept this approach is one of the remaining stumbling blocks for large developing countries like China, India and Brazil from joining the Kyoto *Protocol*. As some of these developing countries are now approaching the emission levels of the developed nations, these issues are becoming even more difficult to resolve politically.

As we have seen, there are a variety of public policy instruments available, with which many nations have experience, and which have been successful in managing an array of environmental problems. These instruments generally have many flexible options available and can be combined with each other or other available business or economic solutions to address Sustainable Scale challenges.

While there are costs involved in applying these solutions, there are also financial and other benefits. It is not the application of the policy instruments themselves, but the focusing on reducing throughput below regeneration rates, that is critical for achieving Sustainable Scale. Applying these public policy solutions to Sustainable Scale challenges is a matter of political will[49].

Institutional solutions

National governments are responsible for protecting their citizens and their environment and for working collaboratively with other governments to deal with issues that cross borders. Sustainable Scale issues are transboundary, and thus require collaboration among affected nations. Solutions to Sustainable Scale problems require determination of optimal scale for a variety of material and energy throughputs that currently challenge the ability of ecosystems to continue providing life support services, and which are organised in ways which currently increase injustices. Determining optimal scale requires broad participation, especially of those who are most adversely affected by the current injustices. Such participation represents a major social challenge as even our best democratic institutions are skewed toward the rich and powerful, who generally have a vested interest in, and undo influence in maintaining, the status quo. Nonetheless, there are many openings within the world's democratic systems for public input and participation. Massive efforts in public education are needed to raise awareness of the importance and pervasiveness of scale problems and to identify possible directions for solutions. Governments in democratic nations remain sensitive to public opinion and changes are likely necessary before governments take the issues of Sustainable Scale seriously.

---

[49] "Trading in Pollution" OECD Observer. August 2002. http://www.oecdobserver.org/news/fullstory.php/aid/750/Trading_in_pollution.html

Taking Sustainable Scale issues seriously would involve both changes in national policies and priorities, and working collaboratively with other nations to ensure attractive solutions are implemented. Ultimately, it is governments and the international agreements they enter and enforce which will determine whether Sustainable Scale is achieved and maintained. But they cannot do it alone and the role of civil society will be key in generating widespread support for governments to implement attractive solutions. For better or for worse, national governments represent critical institutions through which problems of Sustainable Scale must be addressed. Any efforts which assist individual nations to live within their own ecological footprint are positive steps toward Sustainable Scale. But individual nations will not be able to implement effective solutions without working toward global solutions to global scale challenges. Collaboration among nations is essential for this task to succeed.

As the sole body representing all nations and peoples on the planet the United Nations and its many agencies and programs, need to play a key role in implementing attractive solutions to global scale problems. To some extent it is already playing this role vis-à-vis its sponsoring several international treaties such as the *Montreal Protocol* and *Kyoto Protocol*, among many others. However, none of these international agreements has adopted a Sustainable Scale perspective, and each of the current international agreements is lacking from this perspective. It is therefore recommended that all existing UN sponsored environmental, trade and economic treaties undertake a review from a Sustainable Scale perspective to determine how optimal scale might be achieved and maintained.

Some suggestions regarding how a few of these important agreements could be strengthened from a Sustainable Scale perspective have been identified in the Ecological Policy Handbook - Vol. II Areas of concern. However, there are a number of additional areas that require attention, including a massive global public education campaign raising awareness of Sustainable Scale issues and the necessity of transforming the global economic system to achieve and maintain Sustainable Scale. Broad public participation is essential for a variety of reasons. Broad public support will be essential for the radical changes needed to achieve and maintain Sustainable Scale. Creative ideas from all sectors will be required to implement the kinds of changes needed. Most importantly, optimal scale requires consideration of values regarding current and future generations of people as well as for non-human creatures. Articulating such values requires broad participation and development of consensus; they cannot be imposed in a top-down fashion if they are to succeed. The *Global Education 2030 Agenda* of UNESCO, as the United Nations' specialised agency for education, is entrusted to lead and coordinate the *Education 2030 Agenda*, which is part of a global movement to eradicate poverty through *17 Sustainable Development Goals by 2030*. Education, essential to achieve all of these goals, has its own dedicated Goal 4, which aims to *"ensure inclusive and equitable quality education and promote lifelong learning opportunities for all"*. The *Education 2030 Framework for Action* provides guidance for the implementation of this ambitious goal and commitments. As with most UN initiatives, this thematic focus will be implemented through the efforts of individual nations, providing an opportunity for civil society to participate.

Abundant, cheap energy is coming to an end and there are no acceptable substitutes to readily take its place. Abundant, cheap energy, regardless of the source, represents a potential threat to Sustainable Scale. Energy and economic development are inextricably intertwined. For individual nations or international agencies such as the World Bank, International Monetary Fund and the World Trade Organization, to pursue development without a global energy plan is a recipe for continued development failures, social inequities, violent conflicts and ecological disasters. The current political climate may not welcome facing the negotiations for an international agreement regarding energy, but this does not mitigate the need for such an agreement. The more nations that prepare themselves for such a discussion and agreement, the easier it will be when the global political climate more clearly recognises the necessity.

International institutions are created and controlled by individual nations. Several nations have taken the lead in recommending reforms for existing environmental institutions. There is widespread discussion about the inadequacies of existing international environmental agencies and agreements: existing agencies, such as the UNEP, UNSD, the Commission on Sustainable Development, and others, are acknowledged to be under funded, weak relative to trade related agencies such as the WTO, and fragmented. Various proposals are currently under discussion to strengthen global environmental protection, including consideration of a World Environment and Development Organization. Calls for a strengthened environmental agency have even come from the WTO. Potential conflicts exist between provisions of the WTO, which encourage trade and economic growth, and the possible goals of a World Environmental Organization, which presumably would seek to limit the negative impact of economic growth on critical life support systems. Currently, the WTO is well funded, has powerful enforcement mechanisms, and its rules dominate other international agreements. But this predominance of WTO rules is being challenged and opportunities exist for substantial progress. It is encouraging that the global and unprecedented nature of environmental problems is being acknowledged within the UN framework, and that new institutions are being considered to deal with these challenges. Whatever improvements to international environmental governance occur, it is vitally important that a Sustainable Scale perspective is adopted as a priority approach for this new endeavour.

A Sustainable Scale perspective requires not only that ecological sustainability be achieved, but also that issues of social justice are addressed for sustainability to occur. Both ecological sustainability and social justice are critical frameworks for reforming global environmental policies and practices, as well as economic development activities. For the goal of Sustainable Scale to be realised, the dominance of WTO rules which encourage and support unlimited economic growth will have to end. This will be difficult if not impossible unless there is widespread support for placing a higher priority on ecological sustainability and social justice. This emergent perspective, in turn, requires much broader understanding of the interconnectedness of ecological sustainability, social justice and economic development, and the unique role each has to play in contributing to human happiness and well-being.

Chapter IV
Economics for community

Economics is about meeting human needs and, at its core, it is about the exchange of goods and services. It is a basic human activity and occurred long before there were economists and economic theories. Exchanges occurred because they benefited both parties, and did so without advertising, stock exchanges or global corporations. Exchanges of tools and raw materials between regions occurred even before agriculture emerged. Exchanges contributed to survival for some groups, as well as to "luxuries," enriching life both literally and figuratively[50]. Modern economic activities are of course much more complex and formalised, but the basic activity is simple, essential and virtually universal. As human societies evolved and became more complex and more goods and services were available, the rules and roles involved became more specialised and formal. With the differentiation in roles, power disparities developed and the benefits of exchange became more disproportionate within groups. Disproportionate benefits also occurred between regions, but more so because some groups enjoyed advantages conferred by the environmental resources available to them rather than because of innate superior intelligence or abilities, or special relations with the Deity[51]. The disproportionate benefits accrued to those with more power both within and between groups and, of course, those with the power made the rules further ensuring their disproportionate accumulation of benefits. This enduring self ser-

---

[50] Diamond, Jared. Collapse: How Societies Choose to Fail or Succeed. New York: Viking, 2005

[51] Diamond, Jared. Guns, Germs, and Steel. New York: W.W. Norton, 1999

ving dynamic has resulted in a modern economy characterised by enormous disparities of wealth between communities and nations, even as economic concepts and practices become more formal and complex. Indeed, the theories and practitioners of economics played a significant role in generating and maintaining these discrepancies. While modern economic theory purports to be a science, it could just as accurately be described as a socio-political justification for ensuring the most benefits accrue to those with the most power. Both descriptions would be partly right and partly wrong. Critical analyses of current neoclassical economic theory and practice identify both its weaknesses, and point to alternative conceptual frameworks which build on its strengths[52]. Economic activities confer many benefits on people around the world. Many aspects of classical economic theory contributed to the increased production and consumption of goods which conferred these benefits. However, there are also many aspects of neoclassical economic theory and practice that work against community, in terms of the just distribution or sharing of the economy's benefits, and the ecological sustainability of its operation. Sustainable Scale is about economics for community, that is for *"common unity"*. This involves economic theories and policies which explicitly seek to preserve and enhance this common unity in terms of ecological sustainability and just distribution.

Neoclassical economic theory is based on the view that people are insatiably acquisitive and that if individuals act in their own self interest to satisfy these needs, then the greatest good will occur for the largest number of people. This view of people as *"homo economi-*

---

[52] Gowdy, J., C. Hall, K. Klitgaard and L., Krall (in preparation). "The Twelve Most Important Myths of Neoclassical Economics."

cus" is at the core of the idea that the market can provide the greatest good; it is the invisible hand of the market, through individuals acting in their own self interests, that creates the greatest social good. This view of human nature is in sharp contrast with the view that humans are essentially social beings, whose happiness and well being are largely determined by their relationships with each other. The notion of *economics for community* is based on a more holistic and scientific basis of the determinants of human well being. This view recognises that market goods play an important but limited role, and that the non-market goods and services provided by community or nature are also essential factors. These very different views of human nature have enormous implications for our economy, our businesses, our communities, and our survival.

Market goods are those that are exchanged in barter or trade; *non-market goods and services* are those that are not traded or bartered but are available either because they are provided by nature or community. All manufactured goods are market goods. Examples of non-market goods or services are things like community safety, care provided by friends and family, protection from UV radiation provided by the atmospheric ozone layer, and climate stability. Every individual's total welfare is a combination of both market and non-market goods and services: both are essential for human well being. Unfortunately, neoclassical economic's search for continued growth seeks to include ever more non-market goods and services, wherever possible, into the market. Profits can only be made on market goods and services, so the more things that can be included in the market economy, the more profit is available. Once goods become market goods there is a tendency to exploit them as rapidly as possible to maximise profits, and move on and do the same with the next hot market item. In this way various renewable, as well as non-renewable resources are being depleted. Economics for community holds that the non-market goods and services have intrinsic, non-monetary values that cannot be replaced by financial assets. It holds that rules are needed to respect this basic requirement for various non-market goods and services to ensure economic activities serve these broader values. Most ecosystem services are non-market services and need to be protected to ensure these services are sustainable. Ideally, governments are responsible for ensuring rules are in place to protect both market and non-market goods and services. Economic interests often distort this ideal and much greater protection is provided to market items.

Non-market goods such as UV protection and democracy have intrinsic value beyond the monetary. These values have to do with ideals of justice, safety and security, aesthetic preferences and well being. However, many of these non-market goods and services can also be shown to also have monetary value, and various attempts have been made to identify these in dollar terms[53]. There are a variety of methods for assessing the monetary value of non-market goods and services, and the exercise can be instructive to illustrate that these monetary values can be significant (e.g., the dollar value of global ecosystem services was identified to be more than twice that of the global money economy). However, it is important to accept that the essential non-market values of these goods and services are non-monetary and do not require monetary justification to be preserved.

---

[53] Costanza, Robert, R. d'Arge, R. de Groot, S. Farber, M. Grasso, B. Hannon, K. Limburg, S. Naeem, R. O'Neill, J. Paruelo., R. Raskin, P. Sutton and M. van den Belt. "The Value of the World's Ecosystem Services and Natural Capital," *Nature* 387.256 (1997): 253-260.

From a Sustainable Scale perspective, neoclassical economic's focus on continuous growth is a major cause of scale problems. Macroeconomic theory has no *"when to stop"* rule, no concept or procedure to determine when economic growth is producing more costs than benefits[54]. Microeconomics does have such a rule; it is determined solely by market prices as they are affected by supply and demand. A macroeconomic *"when to stop"* rule that preserves ecological sustainability is needed to ensure economic growth does not exceed Sustainable Scale; this can only be determined by ensuring throughput does not exceed regeneration, both for individual categories of throughputs, and for economic throughput as a whole. A macroeconomic *"when to stop"* rule cannot be determined by market pricing, and thus cannot solely be the result of a series of microeconomic decisions. Given the current extent of global trade, *"when to stop"* rules for specific sectors (e.g., forestry, fisheries, mining, energy, etc.) will need to involve global agreements both about the absolute levels of throughput, and the distribution among nations. Such rules would also have to consider both market and non-market items. Clearly, more than economic policies are required to achieve the goal of Sustainable Scale. But current economic policies are responsible for exceeding unSustainable Scale and changes in many of these policies and practices are essential. A variety of suggestions are provided below.

Macro-allocation of market and non-market goods and services

Neoclassical theory and practice seek to include virtually all goods and services into the market economy. The more that can be included, the more the market can grow. This creates serious diffi-

---

[54] Roodman, David. "Getting the Signals Right: Tax Reform to Protect the Environment and the Economy." *Worldwatch Paper* (134) May 1997.

culties for Sustainable Scale, as many of the goods and services essential for human well being are not market goods or services, but are greatly affected by ever expanding market activities (eg. atmospheric ozone layer, global climate stability, etc. Economics for community recognises the existence of non-market goods and the importance of preserving them and protecting them from legitimate market activities. Often, legitimate market activities at one level have significant negative non-market impacts at other levels (e.g., clearing of forests for agriculture may make sense for the local farmer, but have negative consequences for regional hydrologic cycles and global climate). Economics for the global community requires that these non-market externalities are internalised; that theories and practices ensure ecological sustainability and just distribution in the optimal, macro-allocation of market and non-market goods[55].

Governments generally recognise the importance of non-market goods and tend to provide them in the form of public goods — health care, education, roads, streetlights, parklands, etc. Monetary policies are generally not helpful in the provision of non-market goods. Low interest rates to stimulate investment only works for market goods; even with low interest rates the private sector will not invest in non-market goods as the latter provide no profit opportunities. Investing in non-market goods is left to governments. There are two types of non-market goods governments invest in, which have very different implications for Sustainable Scale. When governments invest in man-made non-market goods (e.g., roads or hospitals) this injects money into the economy which is used to purchase market goods and encourages economic growth. Such government in-

---

[55] Roodman, David. "Getting the Signals Right: Tax Reform to Protect the Environment and the Economy." *Worldwatch Paper* (134) May 1997.

vestment acts like direct investment in market goods and contributes to the throughput that must be accounted for in comparing throughput with regeneration of ecosystem services. Governments also invest in non-market goods which are not man-made, but which protect or restore ecosystem functions (e.g., parklands, wetland or species habitat preservation, etc.). Such investments contribute to maintaining Sustainable Scale by maintaining or increasing the regeneration of natural capital. Governments also have a variety of fiscal policy tools at their disposal to protect and enhance non-market goods relevant to Sustainable Scale.

True cost pricing: Pigouvian taxes and subsidies

Economists have long known that every market product has external social and environmental costs. One of the reasons that economic activities exceed Sustainable Scale is because the costs of unsustainable levels of throughput are not included in the market prices of the throughput, thereby giving a false accounting of the benefits of such activities. Economist A. C. Pigou (1920)[56] promoted the idea of incorporating the external cost into the price of products to accurately reflect the true costs. Pigouvian, ecological or green taxes are all terms for the same attempts at fiscal policies to internalise externalised environmental and social costs. If market activities confer a net benefit on community that are not reflected in the price of the goods sold, then a subsidy can be provided to acknowledge that contribution. Such taxes ensure that only those market activities that generate net benefits to community will endure. If companies have to pay pigouvian taxes for the social and environmental costs of their activity, then they will seek to reduce these costs to

---

[56] Pigou, J. C. The Economics of Welfare, 4th ed. London: Macmillian, 1932 (originally published in 1920)

zero, for the benefit of all. If they receive subsidies for net benefits to community, then they can be encouraged to provide non-market goods that contribute to Sustainable Scale. By implementing a pigouvian tax a government is essentially creating a property right to the environment for the state, using a liability rule and ensuring that specific economic activities are a net benefit to community. Indeed, it might be argued that governments have the responsibility to do just that. A wide variety of pigouvian taxes have been used to reduce pollution and encourage environmental protection. Generally these taxes are designed to be introduced over a number of years to provide companies with time to adjust. If the annual increments are small enough to start, and the government policy is firmly committed to the tax over the long term, it is easier for companies to plan and adapt. Such taxes have been effective in[57] reducing overfishing in New Zealand, reducing water use in Chile, reducing toxic wastes in Germany, increasing recycling of demolition waste in Denmark, reducing heavy metal discharges in the Netherlands, reducing nitrogen use in agriculture in Sweden, reducing nitrogen oxide emissions in Sweden, reducing sulphur emissions in the United States.

A feature that can enhance the attractiveness of a pigouvian tax is for the government to agree to make it "revenue neutral" by reducing the least useful existing tax as the new tax is introduced (e.g., reduce taxes on income or labour which discourages hiring, as a pollution tax is introduced). In this way a broader program of tax reform can be implemented, which taxes "*bad ideas*" (against resources, land use or pollution) and eases "*good ideas*" (employment, and ecosystem services). Several European countries have adopted "eco-

---

[57] Daly, Herman and J. Farley. Ecological Economics: Principles and Applications. Washington: Island Press, 2004

logical tax reform" as part of their employment and environmental strategies.

Ecosystems respond to quantity and quality of material throughput, not prices. Controlling throughput via quotas and bans is more effective than manipulating prices. One of the challenges with relying on pigouvian taxes is the assumption we know what harm is being done by the activity being taxed. With the complex and emergent properties of ecosystem dynamics, these costs cannot always be anticipated. However, where harm is known, taxing it will help reduce it. Additional challenges with pigouvian taxes are that the social and environmental costs of specific market activities can be difficult to identify precisely, and can change over time, requiring constant adjustments to the tax. This is problematic for businesses which need to plan ahead and prefer the simplicity of knowing in advance what their costs will be. Nonetheless, reasonable estimates of true costs are possible and adjustments can be made over time. But serious errors could lead to the costs not being covered by the tax and eventually exceeding Sustainable Scale. Ongoing review is essential. There are two kinds of pigouvian subsidies, each of which have different impacts on Sustainable Scale. If a bonus or subsidy is granted to an industry to not pollute, then the industry might become more profitable and attract new entrants — thereby increasing the overall level of pollution. Subsidies or bonuses to industries which actually restore or increase levels of ecosystem function would not have this effect. Combining pigouvian taxes or subsidies with quotas and or emission trading schemes can both set the absolute throughput limits required and provide the disincentives to ensure the quotas are respected.

It might be argued that the close to $1 trillion dollars spent annually on advertising globally produces a variety of social and environmental costs, and should therefore be subjected to a pigouvian tax. Advertising obviously contributes to economic growth, a major cause of exceeding Sustainable Scale thereby imposing serious costs on current and future generations. Advertising encourages consumption of market goods by providing information about the benefits of such goods. Often this information makes people feel deprived if they do not have the good being advertised. It also ignores the potential costs to the purchasers by not providing full disclosure regarding the consequences of their purchases. Advertising's focus on market goods provides no information about the important benefits of non-market goods, except to the extent advertisers attempt to associate their products with these goods (e.g., using your SUV to enjoy the wilderness). This leaves the public less informed of both the environmental and social costs of what is advertised and the critical services that ecosystems provide. Civil society and governments attempt to take up this slack but generally do not have the financial resources available to compete with the private sector. A pigouvian tax on advertising could provide funding for such public information services. Combined with regulations of full social and environmental cost disclosure for all private sector advertising, such a tax would contribute to the optimal macro-allocation of market and non-market goods.

Price Determined Not Price Determining

From a Sustainable Scale perspective allowing the market to determine prices for goods and services is problematic. Market pricing will not provide a "when to stop" price — a price that indicates that levels of throughput are about to become unsustainable. In fact, just

73

the opposite occurs; as a natural resource becomes more scarce its price increases. Higher prices generally lead to reduced demand. But as long as the resource can be provided for a price the market will bear, than the resource is increasingly depleted — the exact opposite of what is required to prevent an unsustainable level of resource throughput. A major investment house projected a "super spike" of $50-110 for a barrel for oil due to increasing demand and decreasing availability[58]. Clearly, prices cannot be relied on to achieve Sustainable Scale. Rules and regulations that determine a sustainable level of throughput (from both a source and sink perspective) fix the quantity of the resource or emission allowed. This fixed quantity and how it is used will then determine the eventual market price. Such a regulation of pricing is required to achieve and maintain Sustainable Scale. The issue is: what determines price — the market or the quantity deemed sustainable? Unrestrained markets do not set prices that will limit quantities to sustainable levels. Determining prices by regulating quantities is in direct opposition to an unfettered market — the goal of neoclassical economics.

Money systems are a convenience; it is much easier to take an accepted currency to purchase food than to carry three pigs to exchange for several bushels of wheat. Issuing and regulating currency is a public good, facilitating and regulating market exchanges. Providing this service confers benefits on the provider. As the currency system facilitates exchange, the issuer of currency reaps some of the benefits of these exchanges. In the past, this benefit accrued to the sovereign, then to the states which succeeded the kingdom. Today, however, almost all nations have granted the authority to

---

[58] "Is there life after $60/bbl?" GS Global Economic website. Global Viewpoint, 23 March 2005. http://www.gs.com/insight/research/reports/docs/global_viewpoint5_05.pdf

create new money to financial institutions. While the states still impose regulations about how this can and cannot be done, many of the benefits which previously accrued to the state now go to these financial institutions. More importantly, a powerful policy tool has been given up by the state. Seigniorage is the benefit that accrues from issuing currency. When states transferred this right from themselves to financial institutions they gave up two important benefits. The financial benefit from creating money no longer accrues to the state but to the financial institution; the state has foregone these revenues that now go to financial institutions. Taxpayers must make up the difference. Secondly, because money creation is tied to loans, new money creation is tied to the economic growth required to repay the interests on those loans. The money supply is thus determined by the amount of loans made by financial institutions. The state has foregone the power to manage the money supply on the basis of social or natural capital. Monetary policy is locked into promoting economic growth, thereby contributing to ever increasing levels of material throughput, regardless of the degradation of critical natural capital[59]. If nation states were to reclaim the right of seigniorage they could both increase their revenues, and expand the policy levers they have to ensure economic activities maintain an overall level of well being. But well being is more than financial wealth and involves social and natural capital as well. All must be monitored and managed to ensure Sustainable Scale.

Monetary reform which returns seigniorage to national governments is no guarantee that they will operate within Sustainable Scale. Such reform is, however, a necessary if not sufficient policy con-

---

[59] Anielski, Mark. "Fertile Obfuscation: Making money whilst eroding living capital." Presented at 34th Annual Conference of the Canadian Economics Association, June 2-4, 2000: Vancouver, http://www.pembina.org/pdf/publications/fertile.pdf

dition for governments to play a more active role in achieving and maintaining Sustainable Scale. And this is the main problem European Member States are facing after the introduction of the single European currency: the euro.

An other problem, this time common to the whole world is represented by the *Gross Domestic Product* (GDP), widely taken as the most important indicator of national well being. Virtually all governments at all levels seek to increase their GDP but, as a measure of economic activity, this indicator fails to capture the impact of economic activities and other public policies on the accounts of social and natural capital. Worse, increases in GDP can actually contribute to declines in these other accounts which monitor wealth in the broadest sense, including the flow of non-market goods and services that determine human well being and happiness. The GDP is an inadequate measure of wealth because it is limited to total expenditures, regardless of what those expenditures are for, and what their impacts are on social and natural capital stocks and flows. Expenditures for oil tanker spills on pristine coastlines, medical expenses for self-induced lifestyle illnesses and diseases, lawyer's fees for divorce, and home security systems for domestic protection, are all examples of economic activities which contribute to GDP which also degrade social or natural capital. What is true of GDP at a national level is also true of GWP at the global level!

" ... *there is a major flaw in measuring the quality and achievement of life by the total of economic production (GNP/GDP) the total of everything we produce and everything we do for money.*"
John Kenneth Galbraith (1999), one of the economists who developed the GDP.

"… the welfare of a nation can scarcely be inferred from a measurement of national income as defined by the GNP … goals for 'more' growth should specify of what and for what."
Kuznets (1965), one of the economists who developed the GDP.

Various attempts have been made to overcome the shortcomings of the GDP, most notably the *Genuine Progress Indicator*, or GPI. This measure incorporates indicators of social and ecological, as well as economic, well being. It also subtracts the defensive expenditures (e.g., crime, family breakdown and environmental degradation, etc.) normally included in GDP. Other advantages include its method of treating the costs of current activities to future generations, as well as the costs of current income disparities. Comparisons of GDP and GPI in developed economies invariably show increasing divergences over the decades from about 1950, reinforcing the notion that economic growth beyond a certain level does not add to human well being or happiness. Various jurisdictions are exploring use of the GPI in monitoring their well being, and as a policy tool for improvements. Analyses using the GPI are helpful in demonstrating how seemingly positive changes in some areas of importance can actually lead to negative changes in other important areas. Continued work is being done to improve the rigour and comprehensiveness of the GPI, as well as other broader measures of wealth. Other alternatives to expanding the GDP include *"triple bottom line"*[60] reporting, which incorporate economic, environmental and social indicators of well being. In many cases these measures are developed by individual corporations who chose whatever non-financial indi-

---

[60] https://www.investopedia.com/terms/t/triple-bottom-line.asp

cators they wish, and which can change from year to year. A more robust set of triple bottom line indicators are being developed by the *Global Reporting Initiative*, a UN sponsored attempt to develop alternative measures. This attempt involves accountants, business representatives, social scientists, civil society organizations and diplomats. They are attempting to develop rules for reporting on the social and environmental indicators that are as comprehensive, reliable and detailed as the rules for financial reporting[61]. Progress is slow, but the initiative demonstrates the widespread recognition for an indicator of wealth or well being that is not restricted to the financial dimension as is GDP.

One of the biggest limitations of the GPI, GRI and similar measures, is that there is not a conceptual framework to guide what variables are included in the index, nor is there a framework to guide the relative weightings associated with each of the variables included. Currently, each of the multiple variables receives equal weighting; there is no differentiation, for example, between the negative costs of crime and family breakdown on the one hand, and degradation of ecosystems on the other. This equal weighting reflects the developers' acknowledgements that assigning weights to different variables raises the question of whose values are to be reflected in the weightings. The developers of the GPI have not yet addressed the issue of developing consensus regarding values, but recognise the need to do so. Those involved with the GRI, are still struggling with an appropriate value framework to apply. From a Sustainable Scale perspective, achieving ecological sustainability is a critical value. The concept of optimal scale requires socio-political

---

[61] "Global Reporting Initiative." GRI. http://www.globalreporting.org/

consensus of how we organise and manage our economy to remain within the limits of ecological sustainability. Both the GPI and GRI currently lack such an explicit value position which ensures ecological sustainability.

Fair wealth distribution: the growth issue

Under the current economic paradigm, economic growth is equated with increased wellbeing. There are many problems with this simplification, but the main one is that aggregate or total growth does not identify the distribution of economic benefits in society. If one person or one million people received the entire economic output, the total GDP would register no difference. Since growth is automatically equated with increased well-being and as the solution to poverty, there is always a bias toward growth. If Sustainable Scale was exceeded, then GDP would register nothing, except a further increase in economic growth. If increased well-being is the goal of economics, than distribution of wealth becomes much more important than growth, and in fact a much more direct way to address poverty. It separates poverty reduction from growth as two separate and somewhat unrelated issues. At the subsistence level there is evidence of a correlation between economic growth and poverty reduction. Beyond that level, the correlation weakens and is more correlated with government policies. If we are to respect Sustainable Scale, then there will come a point at which no further economic growth as measured by material throughput is possible. At this point it will be impossible to continue the attempt to use economic growth as a solution to poverty. Then wealth distribution becomes a much more important issue than economic growth.

The increase in GWP has been fairly steady over the last several decades; those who hold up the GDP or GWP as the key measure of wealth and success point to this phenomenon as an indication that things are improving. They take it as an endorsement of current economic development practices, and call for more of the same. However, measuring the total amount of financial activities in an economy, as does the GDP, is only one way of determining if there is enough money available (putting aside for the moment the non-monetary determinants of wealth). Also of interest is just how those financial resources are distributed within a community, or a nation, or globally. The *multidecadel* increase in GWP has been accompanied by an ever widening gap in the distribution of wealth — the rich get richer and the poor get poorer. In 2019, the 33rd annual *Forbes* list of the world's billionaires, included 2,153 billionaires with a total net wealth of $8.7 trillion, down 55 members and $400 billion from 2018 billionaires.

Economists have a way of measuring the distribution of financial wealth called the GINI coefficient. It ranges from a low of 0.0, where everyone would theoretically have the same amount of financial wealth, to a maximum of 1.0, where all financial wealth would be owned by a single party. In most developed countries the GINI has been steadily increasing over the past several decades, indicating ever greater concentrations of financial wealth in fewer and fewer hands.In the United States it is over 0.78 and rising[62]. While the financial pie is growing, more and more of it is going to fewer people. Along with the concentration of financial wealth goes the concentration of political influence, disrupting the democratic process. While the injustices of large disparities in financial wealth and political influence are undesirable in and of

[62] Quadrini, V. and J. Rios-Rull. "Understanding the US Distribution of Wealth." *Federal Reserve Bank of Minneapolis Quarterly Review* 21.2 (Spring 1997): 22-36. http://minneapolisfed.org/research/QR/QR2122.pdf

themselves, there is also an impact on activities which affect ecological sustainability. With large disparities, both ends of the financial wealth spectrum contribute to ecosystem degradation. The United States, for example, with only 5% of the global population, consumes approximately 25% of the world's resources annually[63]. And within the US, those with the most financial resources consume a disproportionately high amount. At the same time, those at the low end of the financial spectrum, half the world, nearly 3 billion of the world's poorest live on less that $2 a day also degrade the environment in numerous ways because they have little choice in their struggle for survival[64]. Economic policies which target both a minimum guaranteed income and lower GINI coefficient could contribute to both Sustainable Scale and just distribution.

A guaranteed minimum income ensures that every member of a community has sufficient financial resources to meet basic human needs. When people's needs are met, they are less likely to degrade ecosystems, engage in violence and commit crimes. A guaranteed minimum income can thus provide public benefits as well as personal and environmental ones. There are a variety of economic policies used by different jurisdictions to provide some sort of minimum income, including various welfare programs, unemployment insurance, minimum wages, and negative income taxes for the unemployed. Guaranteed employment at a living wage is also an option. Funding for such initiatives could be made available from a variety of approaches that have not yet been widely used, such as resource trusts or land taxes. Both approaches are based on the notion that

[63] "China and the Final War for Resources." Bill Radley. Energy Bulletin. http://www.energybulletin.net/4301.html

[64] "Poverty Stats and Facts." Causes of Poverty. http://www.globalissues.org/TradeRelated/Facts.asp

natural resources, including land, are owned collectively and the benefits derived from them should be shared collectively. Instead what typically happens is that governments grant rights to resources use that are many times lower than their real value, and generally to very few parties. These few individuals or corporations then exploit the resources for their own gain, rather than allowing the benefits to accrue to the broader community of citizens. A similar dynamic occurs with land. Land is fixed and part of the common heritage of a nation. The market value of land is determined by proximity to others, i.e., by a social factor influenced by government regulations. Taxing land rather than the buildings on the land would return some of the land values to the community. The main point is that there are innovative economic policy solutions available to encourage more just distribution of wealth, a necessary condition for Sustainable Scale.

If there are absolute limits on the amounts of throughput that can be ecologically sustainable, then it follows that there should be also be limits on the amount of income or wealth that can be accumulated by any individual on the basis of this throughput. This notion of an income or wealth cap is controversial. People from both developed and developing countries feel that anyone should be allowed to earn or acquire as much as they legitimately can (some even drop the legitimacy requirement). The idea of an income or wealth cap is based on the idea that it is indeed legitimate to keep wealth that has been earned through one's efforts or abilities, but not to capture wealth created by nature, society or the work of others. There is also the issue of intergenerational justice. If wealth accumulation is unlimited, then less and less will be available for future generations. If consumption must be limited to stay within

Sustainable Scale, then it seems just that those with the greatest wealth should limit their ability to consume. The value of an additional $100 to someone with great wealth is immaterial, while to a poor family it could be the difference between life and death. Accumulation of great wealth is not required to satisfy basic human needs, a comfortable level of sufficiency, well being or happiness. Great wealth does confer power and status, but these are personal and not community benefits. However, the envy that great wealth stimulates often results in others attempting to gain the same personal benefits, and stimulating economic growth to do so. More growth means more throughput, and increases the likelihood of exceeding Sustainable Scale. A greater public good accrues if wealth is capped and just distribution and Sustainable Scale are encouraged by economic policies. Economic policies to cap incomes and wealth exist, and to some degree are used in some jurisdictions. Progressive income taxes exist in most countries, where taxes increase with income. If such taxes approached 100% they could be used to cap income. Consumption taxes, such as a VAT (value added tax), also help reduce consumption. A special tax on luxury goods would do the same. A *Material and Energy Throughput* (MET) tax has also been proposed[65], as a way of capturing not just the market value of the good or service purchased, but the throughput value in terms of its impact on ecosystem function. Yet another approach would be to limit the range of salaries between the highest and lowest workers in a firm; a fixed ratio of, for example, 10:1 would ensure that the highest paid employees would have clear limits. Progressive wealth taxes are another option. Currently, almost half of accumulated

---

[65] Paehlke, Robert C. Democracy's Dilemma: Environment, Social Equity and the Global Economy. Cambridge, MA: MIT Press, 2003.

wealth is inherited. The progressive wealth tax on real estate could be expanded to all forms of financial wealth.

Recognition that there is an intimate connection between just distribution and Sustainable Scale is critical to achieving both essential objectives. While there are economic policy options available to achieve both, some of them are not widely known or accepted. There are also attitudinal obstacles to many of the policy options, based on misunderstanding of either the reality of biophysical limits, or the relationship between perceived personal freedoms and public goods. Given the urgency of achieving Sustainable Scale, these obstacles are challenges to be overcome.

Discounting policies

Minimum incomes and wealth caps are means of increasing just distribution largely with the current population that shares the planet. Economic policies also affect future generations. Economic practices which consider the future costs of decisions made today apply a discounting approach to these future costs. The value of money today will be worth less in the future; the discounting rate adjusts present and future values to allow comparisons so decisions can be made which best allocate resources between the present and the future. Generally the discount rate is at or near the interest rate. This practice makes good sense for individual decisions, and over relatively short periods up to a few years. For very long term decisions, and especially for ones that involve non-market goods and services, there are serious problems with the standard discounting approach. First of all, conventional discounting practice assumes that the way people discount the future is exponentially. If this were true then different values would be assigned to costs 100 or 110 years from no. However, empirical studies show that people give more weight to what happens in the near future, but are indifferent to outcomes over very long periods of time. Therefore, applying conventional discounting practices to non-market goods and services, such as climate change, places less and less value on long term future costs, although this does not reflect the way people actually value the future. The lesson for Sustainable Scale is that conventional discounting practices should not be applied to the non-market goods and services provided by ecosystems over very long time spans. But it is those long time spans that are relevant to Sustainable Scale. Whether improved discounting techniques, or totally different ways of assessing future value are needed, is an open question. For example, hyperbolic (vs. exponential) discounting is said to more accurately reflects actual human preferences. Given that certain ecosystem functions are essential and irreplaceable to human well being, our task is to preserve these essential services, and develop economic theories and practices that inform us how to best do so. Basing policy decisions on conventional techniques that lead to destroying life supports for future generations is equivalent to condemning our descendants to an increasingly miserable existence.

Economic Globalization

Economic globalization is the primary policy objective of most developed and many developing nations, as well as the World Trade Organization. The avowed aim is to integrate all national economies into a single global market, regulated by the same rules, including the reduction or elimination of trade barriers, and the inclusion of whatever non-market goods and services are convertible to market items. Economic globalization is said to increase efficiency and bring gains to all parties. Many of the assumptions of economic globalization are questionable, especially from a social justice perspective. To the extent economic globalization increases wealth disparities between nations, it also contributes to injustice, and at least indirectly to Sustainable Scale problems. Our purpose here is primarily to focus on its direct contribution to problems with exceeding Sustainable Scale. Economic globalization is based on the *"Washington consensus"* model of development, also called TINA (*There Is No Alternative*) by former British Prime Minister Margaret Thatcher.

The Washington consensus (TINA) model of development is based on the following factors which are designed to increase the country's GDP or throughput of goods and services: *(i)* deregulation/privatisation, *(ii)* free movement of capital, *(iii)* free movement of goods (No trade barriers), *(iv)* develop products for export, *(v)* reduce public expenditures (structural adjustment).

Increased GWP is an explicit goal of economic globalization. From a strictly economic perspective, increased GWP is desirable as it means economic growth and increased profits for those who make it happen. However, increased GWP also means increased material and energy throughput, which is the key issue in determining whether Sustainable Scale is exceeded. International trade has become a vehicle for the dominant policy goal of ever increasing economic growth. This goal is misplaced as it is in direct contradiction with basic physical laws. Trade has existed between different peoples from the earliest records of human history. Goods were exchanged because they were not available locally. Over time, classical economic theory articulated the notion of comparative advantage, identifying the conditions under which nations could trade to their mutual advantage, including even those goods that were available locally. However, economic globalization has violated several of those conditions, drawing into question the conceptual basis for its justification. Where multiple competing firms were a precondition, we now have a few firms dominating the market; where having financial capital remain national was necessary, we now have global financing; where a moral framework was expected to ensure economic activities had social constraints, we now have an unfettered market where profit is the dominant value. The result is increasing disparities between nations, and increasing competition for increasingly scarce resources; more profits, more throughput and more unsustainable practices. International trade which allows nations to efficiently supplement whatever their unique resource requirements might be to maintain an optimal population within a sustainable range is a possible alternative to international trade for the sake of profit. Each country has different needs to supplement its natural resources to maintain a reasonable quality of life within a Sustainable Scale. This assumes that each nation is attempting to manage its affairs to remain within Sustainable Scale, and only importing or exporting those items which allow it to do so.

Today's trade regime is quite different. Markets, not sustainability criteria, determine what is imported or exported. Financial considerations trump ecological and justice issues. Resources are exported because they generate revenue, and goods are imported because they are "cheaper" than produced locally. The privatisation schemes favoured by global traders often involve the transfer of a public good (e.g., water supply) from government to private sector control.

The result is liquidation of national accounts of natural capital, destruction of communities as their industries are closed due to foreign competition, and social unrest resulting from a public good now being treated as an ordinary market commodity. Exporting non-renewable resources often creates considerable ecological costs for the exporting country. Relying on non-renewable resources is economically, as well as ecologically unsustainable. Exporting renewable resources only makes sense if the exporting nation has an ecological surplus of the resource. Unfortunately, this is not always the case. Even when the resource in question is not needed by the exporting nation, the externalised costs of environmental and social damage may outweigh the financial benefits of the trade. Importing non-renewable resources is also short sighted; the consequences of dealing with their ultimate depletion can be immense, and to import renewable resources that are not surplus to the exporting nation's sustainable needs is only to shorten the time over which these resources will remain renewable.

International trade is essential to a just and sustainable planet, but not if it is based solely on profit and promoting growth. Basing international trade on sustainability and justice criteria is as important as basing national economies on these same values.

Chapter V
Sustainable business practices

Economic structures and policies create the context for through-put, but it is individual business operations that are responsible for the actual throughput activities in the production of goods and services. The specific technologies they use have a significant impact on the amount and types of materials that go through the production cycle of extraction, manufacture, use and waste. In the past, the impact of various technologies on the environment was not a significant issue for business planning and operation. The effects of burning coals were too obvious to ignore in the late XIX and early XX centuries, but even British royalty could not exert enough authority to reduce the pollution created. Concerns for the environmental impact of business activities arouse in earnest in the latter part of the XX century with the evidence of chemical pollution, a variety of environmentally significant industrial accidents, opposition of indigenous peoples to logging and mining operations, and the discovery of human impacts on global systems such as the atmospheric ozone layer and climate. The accumulating evidence of serious environmental and human health effects triggered citizen protests and led governments to begin exerting influence on businesses to clean up their act. Environmental concerns arose for individual nations (e.g., with the US establishing the *Environmental Protection Agency* in 1969) and within the United Nations (e.g., UNESCO's *Intergovernmental Conference for Rational Use and Conservation of the Biosphere*, in 1968). The evolution of environmental non-governmental organizations (NGOs) also played a significant role in persuading businesses to take greater care of the environment (e.g., the *Environ-*

mental Defense Fund began in 1967, and *Greenpeace* emerged in 1972). Pushed by these external forces, businesses in the last quarter of the XX century began to take up environmental issues in a more serious way. Here, we try to identify some of the major trends within what is broadly known as the sustainable business development movement and critiques these activities from a scale perspective.

Concern for environmental issues was also connected to worries about the growth of the human population. Projections of a global population of several billion raised doubts about how to provide for such numbers and how to close the poverty gap. It become evident that affluent lifestyles would not be possible for the world's billions, and that some radical changes were needed to both provide for humanity's needs and protect the environment. In the early 70's the Club of Rome published a controversial report *"The Limits to Growth"* which concluded that economic growth could not continue without disastrous environmental and social consequences. Warnings about limits became a hot political topic but were too controversial to stimulate directly relevant government intervention. Instead the more acceptable goal of environmental protection was adopted. Later, the term *"sustainable development"* was introduced by the UN sponsored *World Commission on Environment and Development*. The term has taken on many meanings depending whether the focus is on sustainability or development. Considerable care is needed in understanding how the term is used, as different uses can actually have diametrically opposed meanings.

Sustainable Business Development

Broadly speaking, sustainable business development involves the application of sustainability principles to business operations. Sustainability in this sense can mean a variety of things — ecological sustainability, social sustainability or even sustained economic growth. As such, the sustainable business movement is a component of the broader movement toward greater corporate social responsibility. Interests in this area is reflected in the growing number of business organizations exploring these issues, the large number of related websites, journal and book publications, academic programs in business schools and other faculties (e.g., engineering), standard setting organizations (such as the ISO 14001 standard for corporate environmental management systems), and government initiatives of various types. Even the financial sector is involved, by establishing standards for lending criteria regarding environmental protection and sustainable development, by developing indices for sustainable business practices (e.g., *Dow Jones World Sustainability Index*), and developing environmentally oriented investment opportunities. The overall impact of these efforts in terms of ecological sustainability is still small. But the presence and vigour of this movement is a vital, if not sufficient, component of the need for scale relevant policies and practices. Let's see what the results have been in this sector and what the indications are necessary to face the problems of scale.

Eco-efficiency is the term used by some businesses to describe their goal with respect to the environment. The *World Business Council on Sustainable Development*, an association of some of the world's largest corporations, defines eco-efficiency *"as being achieved by the delivery of competitively priced goods and services that satisfy human needs and bring quality of life, while progressively reducing ecological impacts and resource intensity throughout the life cycle, to a level at least in line with the Earth's estimated carrying capacity"*[66]. This definition is impressive in its reference to human needs and quality of life, which could be interpreted as a move

---

[66] "World Resource Institute." http://climate.wri.org/

away from the mere acquisition of material goods for their own sake and an emphasis on their contribution to meeting needs in the service of human well-being. This is also an impressive definition from a scale perspective in its reference to carrying capacity. This implies recognition of limits, the core of the scale perspective. Businesses from different associations and various sectors have used a similar set of approaches to achieve their own version of sustainable business practices. Much of the improvements in business practices have come from new ways of thinking about meeting customer needs, and redesigning production operations with environmental concerns in mind. Such procedures as life-cycle analysis, design for environment and preventive engineering[67] have all played significant roles in assisting businesses in moving toward more sustainable operations. This trend reflects the call for prevention rather than relying solely on rehabilitation or *end of pipe* solutions. Some of the major procedures include

1. *Increasing of Resource Productivity* to getting more goods or services from less material or energy; it is sometimes referred to as dematerialisation. In this case, knowledge flows are substituted for material flows and there are a wide range of examples where this approach has been successfully applied:

   • carpet manufacturers have reduced energy, water and waste flows from 35 to 89%;

   • redesigning windows resulted in enhancing daylight by 600% and reducing solar heat penetration by 400%;

---

[67] Preventive engineering approaches make use of information on how technology affects human life, society, and the biosphere, so as to adjust engineering theory and practice to create a greater compatibility between technology and its contexts.

- steel pylons for electrical transmission lines use only one third the material required by competing concrete pylons, and they last twice as long;
- making steel with electric smelting techniques uses far less materials and energy compared to more traditional basic oxygen steel production;
- reduction of the total pounds of material used annually while increasing shareholder value per pound;
- reduction of virgin material that goes into carpet manufacturing accepting recycled carpet, which would otherwise have gone into landfills.

2. *Items Large and Small* applicable to all business operations and go beyond the traditional "green businesses" of recycling and waste management. Everything from razor blades and ball-point pens are being made with less materials and resources than they were in the past. Even the ways large buildings are being designed and constructed are affected by this approach. The traditional way a large complex building was designed involved many technical professions adding their input to a set of plans that would be passed from one group to another in the right sequence. Each professional group would provide their design input and pass the plans on to the next group. Typically the method of reimbursing this design input involved compensation based on capital expenditures, therefore, the larger the air conditioning unit for example, the larger the fee. This approach creates few incentives for any group to reduce either construction or operating costs, or to make the building environmentally friendly. An innovative total systems approach to buildings pays special attention not only to the types of materials that are used

(low waste, non-polluting, local origin if possible, etc.) but also to the design process itself. By creating a design team consisting of the various professional and technical groups together with those who will use the building, and setting a goal of making it as environmentally friendly as possible, can lead to dramatic results in terms of reducing material and energy needs. When the professional compensation is based on the reduction in the building's life-long operating costs, their expertise is redirected to saving energy and materials as much as possible. One of the reasons that businesses have been attracted to increased resource productivity is that these redesign investments can save money. The less energy and materials used, the lower their operating costs. Many of the examples in this sector have produced investment returns in relatively short periods. Another reason dematerialisation is attractive to many businesses is that it results in less waste, another goal of sustainable business development.

3. *Waste Reduction* from a business operation can be problematic in several ways. Disposal can be costly more so if the waste is hazardous. If an industrial process is wasteful, it also means input costs will be higher. Many companies are beginning to view waste as a measure of inefficiency and several have set a goal of zero waste for their operations. This approach is also referred to as closed loop production, whereby output from one operation becomes input for another. Remanufacturing is a process several businesses are using to reduce waste and get more productivity from what they produce.

4. With the *Transition to Renewables,* current energy technologies are creating serious environmental problems and it is possible that

global energy supply will not be able to keep pace with anticipated demand. The world is beginning to experience the end of the fossil fuel era, and a transition to renewable energy sources is needed. Businesses have responded by reducing their dependence on fossil fuels and increasing their use of solar, wind and other renewable sources of energy. Several of the major oil companies are becoming producers of renewable energy. Research and development is contributing to solar and wind turbine technologies becoming increasingly efficient. Experience has shown that dramatic reductions in greenhouse gases and other pollutants can be achieved with simple efficiency improvements such as cogeneration and the use of wastes from agriculture and landfills. We are in the early stages of a transition to renewable energy sources. Businesses are preparing both by exploring the direct use of renewables, as well as by reducing their energy needs. With such dramatic reductions in energy demand, it becomes feasible to consider new plant designs that incorporate on site renewable energy sources to power the operation.

5. *Extended Producer Responsibility* is another green procedure. In the past, when goods were manufactured which had toxic components, the toxicity was a problem for those who purchased the products, or those affected by their disposal. That has changed with acceptance of the notion *"if you produce it, you own it... forever"*. This extension of the producer's responsibility beyond the sale of the product is having profound changes on how businesses provide goods and services. With the idea of extended producer responsibility, businesses become much more interested in the design of their goods, because once the useful life of the

goods is over, the goods come back to the producer. This increases the producers' interests in designing goods that are either biodegradable or recyclable. It encourages designs which are more durable, which allow the goods to be easily disassembled, and once disassembled to be reused or composted, leaving no toxic residue to deal with. What was formerly waste and a liability now becomes an asset — a source for new product materials! This is another example of closed loop production, where attempts are made to obtain the maximum productivity from materials over repeated loops through the production cycle.

6. *Imitating Nature* approach, introduced another term which describes the closed loop process that includes both waste production and extended producer responsibility: *"biomimicry"*[68]. This approach emphasises that nature offers many tested examples for businesses to learn about sustainability, for whatever survives in nature has undergone millions of years of trial and error, and is the most efficient solution possible. This can be applied to developing new products, or processes. Velcro is perhaps the most well know example of biomimicry. Many opportunities remain. We have yet to understand how a spider can produce a fibre stronger than steel without using high temperatures, toxic materials or high pressure. Abalone shells provide a lightweight but fracture-resistance crystalline coating, without high temperatures or pressures, which we cannot yet duplicate. There are also many process applications ranging from new designs for computers based on properties of DNA molecules, to filtration systems based on cell membranes, to the layered construction processes used in deer antlers now being applied in nanotech-

---

[68] www.biomimicry.org

96

nologies. This line of thinking has created new fields of study such as industrial ecology, in which sustainability principles are applied to how various industries are sited in relation to each other to make the most of all resources available. Several ecological industrial parks have been built where the traditional waste from one industry is used as input to another, imitating natural processes in a mature ecosystem where nothing is wasted. Physical proximity is not necessary for sustainable business associations to promote sharing of information and strategies not only among members, but also to individual businesses around the world.

7. *Being Green by Buying Green.* Procurement programs are contributing to the popularity of green products and services. Green procurement involves identifying and giving priority to green products and services in corporate and government purchasing decisions. Governments in particular tend to be large volume purchasers and by giving preferential treatment to green products and services, assist their market penetration. Because products and services designed on sustainability principles are still new and exceptional, it can often be difficult to find those pumps, motors or fans, for example, which are most energy efficient or manufactured without using environmentally damaging processes. Green procurement programs facilitate such information gathering and ensure that the purchases made do the job.

8. *Green Certification.* One of the aids to public and private sector green procurement programs, as well as to individual consumers, are a variety of green certification programs. These programs generally focus on a specific industry or sector (such as forestry or fisheries, for example) and establish standards for best practi-

ces from an environmental perspective. These standards are then applied to particular companies in that sector, and if standards are met, the company or its products gets a certificate of approval attesting to its environmental practices. These certification programs were initiated by NGOs wanting to put pressure on companies to improve their environmental performance. In some cases, industries set up their own certification programs to preempt independent reviews. The number and type of green certification programs are spreading and cover such diverse areas as forestry, fisheries, tourism, restaurants, bananas, coffee, and even automobile junk yards. The smart phone wars have led to heated discussions over standard setting in technology markets. It seems only a question of time before the standard setting debate spills over into other areas. An average consumer comes face to face with product certification in her local grocery store. Organic meat, cage-free eggs and fair-trade coffee are only a few examples of certified products, which increasingly populate store shelves. The expansion of green markets has also prompted certification and labelling programs for a wide range of products (e.g., natural, recyclable, eco-friendly, low energy, recycled content, non-toxic, etc.). Obviously, the usefulness of certification is not limited to food or eco-friendly market segments. Typically, certification comes into play in industries where the important characteristics of products are difficult to tell after the purchase has been made. We can't verify where the tomato was grown or whether a bulb is really energy efficient. In other words, since the consumer cannot verify the truthfulness of each manufacturer's claims, the goal of product certification is to bridge the information gap. Green certification is an important step in moving to a more sustainable economy, but much remains to be done[69]. More rigorous standards are needed with respect to sustainable ecological scale; and certification standards need to become the norm, either through legislation or strong social or market incentives. Currently, the onus is on the purchaser to seek out the environmentally friendly product or service; this requires an additional effort and is therefore a disincentive. But as with many aspects of sustainable business practices, these problems are typical of early stage developments of a major transition, and are best viewed as challenges to be overcome rather than solely as frustrating obstacles.

The sustainable business development movement is one aspect of the growing interest in corporate social responsibility. Some individual corporations have developed their own unique triple bottom line (covering financial, environmental and social performance), or supplementary environmental reporting formats. The rigour and quality of these reports is highly varied. Some are very thorough and admit to ongoing problems as well as record successes in environmental improvements. Without explicit standards, corporations can chose what they report and the report format may change from year to year, allowing the corporation to hide certain aspects of their performance record. Some companies have joined with others in their industry or region to collectively set standards and agree on reporting formats. Several dozen groups have emerged that articulate shared standards and work collectively to provide adequate performance records. One of the most comprehensive and rigorous is

---

[69] http://competitionlawblog.kluwercompetitionlaw.com/2013/03/14/product-certification-the-next-big-standard-setting-debate/

the *Global Reporting Initiative*[70], developed in 1997 by the *Coalition for Environmentally Responsible Economies* (CERES) and the UNEP[71]. It involves a long term, multi-stakeholder process with inputs from NGOs, governments, business groups and professional accountants. The goal is to develop corporate reporting guidelines which include reports on social and environmental performance which has at least the same rigour and comprehensiveness as the standards for corporate financial reporting.

The sustainable business development movement is an important step toward a sustainable economy. Wide spread adoption of the approaches described above could mean significant reductions in the use of non-renewable virgin resources, a transition to renewables, elimination of toxins, and a reduction in waste. In addition, unions are supportive because more jobs are generally created as the company expands the services it provides. These examples show that reduced throughput is possible while still meeting human needs, and profits and jobs can be retained in the process. Despite its demonstrated successes to date, the contributions of this movement have been small. As stated by the *World Business Council on Sustainable Development*, "*The troubling news is that it is not being tried on a large enough scale, even though it makes good business sense.*" There are a number of additional shortcomings that need attention.

From a scale perspective one of the most important issues is the setting of limits for environmental impacts. This is rarely done by individual business operations. When a company achieves a target of reduced emissions or improved productivity, as significant and

---

[70] https://www.globalreporting.org/Pages/default.aspx

[71] Global Reporting Initiative. UN Environmental Programme. www.uneptie.org/outreach/reporting/gri.htm

important as this may be, it does not necessarily mean a reduction in absolute levels of emissions or material use. Back in 1865 Jevons wrote about how increased efficiencies in the coal industry actually led to more uses of coal. The same phenomena often occur with achievements in the sustainable business movement. Without first establishing the acceptable scale boundaries which all companies have to operate within, increased efficiencies and resource productivity alone will not ensure ecologically Sustainable Scale. As Jevons noted, such "successes" could actually make the situation worse, and increase absolute throughput and ecological impact. There is the added phenomenon that these "successes," impressive as they are, lull us into believing we actually are solving the problem and prevent us from addressing the issue of scale. In the end, nature does not respond to "miles per gallon" but to "gallons."

Clearly setting scale limits is the most serious challenge to the sustainable business movement actually becoming ecologically sustainable. However, relying on individual corporations to accept this goal, and move with all due speed to accomplish it, is a risky and uncertain route to ecological sustainability. Without clear scale targets, there is no way of determining whether the successes of the sustainable business movement are bringing global economic impact near these targets, or how far we remain from them. Another difficulty regarding the sustainable business movement is that it remains a voluntary process. Those corporations committed to sustainable practices are to be congratulated and supported for their pioneering efforts. They have demonstrated what is possible and point the way to new opportunities. At the same time, we must recognise the obstacles inherent in a voluntary approach. Voluntary approaches leave both producers and consumers to sort out complex and difficult to obtain information about compliance with sustainability standards. This extra effort is an impediment to expanding the market for certified goods and services. It would be easier for corporate and individual purchasers if governments required sustainability standards to be met. This would put the onus on producers to design their operations to meet standards, and governments to ensure compliance.

Governments could ease the transition to sustainability regulations by announcing a commitment to implement such standards over an extended multiyear timeline. This would give corporations an opportunity to plan for the changes required, with the knowledge that change was certain. A second approach would be to initially set relatively easy to achieve targets with respect to limits, and to provide significant incentives for exceeding the targets. Government support for the research and development required to make the necessary transitions would also assist, as would the removal of subsidies that currently encourage clearly unsustainable business practices. It is most important that governments recognise the absolute necessity of achieving ecological sustainability, and carry out the many roles they have of supporting this goal. Governments establish the framework in which individual businesses operate, and this framework will either encourage sustainable or destructive business processes; it will not be neutral.

National governments cannot establish such a framework independently of each other. Sustainable Scale has local and regional implications, to be sure, but these cannot be adequately addressed without first addressing scale at the global level. International agreements are needed to set scale at global limits before individual nations are allocated their share of resources or sinks. This will not be an easy task if experience with the *Kyoto Protocol* provides any indication of the complexities involved. Issues of national sovereignty, identity and competing interests are still prevalent. But it is difficult to envision how ecologically Sustainable Scale can be achieved without such international cooperation. There is no doubt we have the capacity for such cooperation in the face of serious common challenges; the question is whether we will realise this capacity in time to avert social and ecological disaster.

Chapter VI
Lifestyle solutions

Humanity's collective challenge is to ensure our economic and other activities remain within the biophysical limits of the ecosystems upon which we depend for our survival and well being.

But, in which way we should perform such activities?

Sustainable lifestyle is a concept containing a variety of definitions and solutions. All over the world, people are coming together for a brighter and cleaner future, while other institutions have been created to accelerate the shift towards sustainable consumption and production in both developed and developing countries. As far as mentioned, we present the work of two platforms dedicated for the purpose.

Adopted in 2012 at the *World Summit on Sustainable Development*, the *10-Year Framework of Programmes on Sustainable Consumption and Production* (10YFP) is a global commitment to accelerate the shift towards sustainable consumption and production in both developed and developing countries. *Sustainable consumption and production* has been included as a stand-alone goal (SDG 12) of the *2030 Sustainable Development agenda*, and Target 12.1 calls for the implementation of the 10YFP.

The *One Planet network*[72] has formed to implement the commitment of the 10YFP. It is a multi-stakeholder partnership for sustainable development, generating collective impact through its six programmes: *Public Procurement, Buildings and Construction, Tourism, Food Systems, Consumer Information, and Lifestyles and Education.* The One Planet network is an open partnership, and countries in-

---

[72] www.oneearthweb.org/

cluding all relevant stakeholders and organisations are invited to join and actively engage. The strategic objective of the One Planet network over during the period 2018-2022 is to be recognised as the lead mechanism to support and implement the shift to sustainable consumption and production patterns, contributing as an effective implementing mechanism of Goal 12 of the *2030 Agenda for Sustainable Development*. The strategy of the One Planet network for the period 2018-2022, *One Plan for One Planet*, details the common approach around which the entire network can rally.

On September 25th 2015, countries adopted a set of goals to end poverty, protect the planet, and ensure prosperity for all as part of a new sustainable development agenda. Each goal has specific targets to be achieved until 2030. The United Nations Secretary General has highlighted significant gaps regarding SDG 12[73], on "ensuring sustainable consumption and production patterns" which is currently covered in a fragmented and piecemeal way. The ambition and breadth of the *Sustainable Development Goals* make them simply unattainable without robust partnerships. The One Planet network has the legitimacy and mandate from the international community to be the premier multi-stakeholder implementing mechanism for SDG 12, drawing stakeholders together around a common approach and a shared set of objectives. Though a stand-alone goal (SDG 12) has been included, Sustainable Consumption and Production should be seen as an enabler for the implementation of a range of other goals and many of their targets. Achieving sustainable consumption and production will deliver not only SDG 12, but simultaneously contribute significantly to the achievement of

---

[73] General Assembly Economic and Social Council A/72/124–E/2018/3

almost all of the SDGs[74], directly or indirectly. The sectoral and thematic programmes of the One Planet network are a constellation of organisations from sectors and regions around the world, acting as the implementers of the One Planet strategy. Growing portfolios of activities highlight the products and solutions the programmes offer (or are developing) to support countries in the shift to Sustainable Consumption and Production. This platform is where the lead organisations of each programme can coordinate their efforts, to showcase what the programme is doing, who is taking part, and how interested stakeholders can engage.

Sustainable Public Procurement[75]: a vision of a world in which environmental, economic and social aspects of sustainability are embedded in public procurement policies strategies, processes and practices while promoting good governance and integrity in public procurement. Sustainable public procurement it's "a process whereby public organizations meet their needs for goods, services, works and utilities in a way that achieves value for money on a whole life cycle basis. This means generating benefits not only to the organization, but also to society and the economy, whilst significantly reducing negative impacts on the environment". Through SPP, governments can lead by example and deliver key policy objectives and send strong market signals. Sustainable procurement allows governments to reduce greenhouse gas emissions, improve resource efficiency and support recycling. Positive social results include poverty reduction, improved equity and respect for core labor standards. From an economic perspective, SPP can generate income,

---

[74] https://www.resourcepanel.org/reports/assessing-global-resource-use

[75] https://www.oneplanetnetwork.org/sustainable-public-procurement/about

reduce costs, support the transfer of skills and technology and promote innovation by domestic producers.

Sustainable Tourism Programme[76] is a multi-stakeholder partnership that promotes knowledge sharing and networking opportunities to better implement sustainable consumption and production in the tourism sector. The 10YFP *Sustainable Tourism Programme* (STP) envisions a tourism sector that has globally adopted *sustainable consumption and production* (SCP) practices resulting in enhanced environmental and social outcomes and improved economic performance. Its mission is to catalyse a transformation for sustainability, through evidence-based decision making, efficiency, innovation, collaboration among stakeholders, monitoring and the adoption of a life-cycle approach for continuous improvement. Through the development and implementation of activities, projects and good practices in resource efficient and low-carbon tourism, 10YFP STP steers the tourism sector towards enhanced sustainability by reducing the loss of biodiversity, preserving ecosystems and cultural heritage, while advancing poverty alleviation and sustainable livelihood. One of the most significant characteristics of tourism is its transversal nature as an economic sector with multiple links across related industries, elaborate supply chains and multi-stakeholder networks. These can be used to systematically encourage the shift towards the more sustainable development of the sector. Adopting a life cycle approach in tourism design and operations will also engage consumers in actively promoting SCP. Tourism is one of the main economic sectors in the world, accounting for 10% of GDP (direct, indirect and induced), 7% of the world's exports, and one in 10 jobs. International tourist arrivals (overnight visitors) in 2016 grew by

---

[76] https://www.oneplanetnetwork.org/sustainable-tourism/about

3.9% to reach a total of 1,235 million worldwide, an increase of 46 million over the previous year. UNWTO forecasts international tourists to reach 1.8 billion by 2030. SDG 12 focuses on ensuring SCP and includes targets related to the sustainable management and efficient use of natural resources, food loss, waste generation, sustainability reporting of business and the development and implementation of tools to monitor sustainable development impacts of sustainable tourism, among others.

Consumer Information for SCP is a global platform supporting the provision of quality information on goods and services, to engage and assist consumers in sustainable consumption. This takes many forms, from labels on products; to advertising, marketing, and public and third sector awareness-raising campaigns; to communications between peers via social media or family and friend networks. This includes eco-labels, voluntary standards, product declarations, ratings, marketing claims, foot printing, life-cycle assessments, etc., and other ways of communicating with consumers on environmental and social issues connected to products. They may focus on a single issue, or follow a life cycle approach considering the impacts of every stage of the product development process, including how a product is used and how it is treated responsibly at end-of-life. Research indicates that the demand for sustainable goods and services is high and growing, but consumers often remain unable to make informed choices, or simply do not act according to their intentions. Key reasons for this include a lack of transparency, incomplete or unreliable information, and the proliferation of labels and standards, which complicate the comparison of information. The importance of providing reliable information is recognised in target 12.8 of the Sustainable Development Goals: "by 2030, ensure that people everywhere have the relevant information and awareness for sustainable development and lifestyles in harmony with nature".

Sustainable Buildings and Construction[77] has the goal of promoting resource efficiency, mitigation and adaptation efforts, and the shift to SCP patterns in the buildings and construction sector. It was launched in the UN-Habitat Governing Council Side Event on 20 April 2015. The SBC programme will improve knowledge of sustainable construction, develop sustainable solutions, and share that knowledge and those solutions globally. Although an abundance of solutions already exist, there is still plenty of room for more innovations. Through the programme, all major projects in the field of sustainable construction can be brought together under the same umbrella. The goals of the programme are to promote resource efficiency, mitigation and adaptation efforts, and the shift to SCP patterns in the buildings and construction sector. These goals are being operationalised through several concrete work areas: *(i)* foster enabling frameworks to implement SBC policies; *(ii)* promote Sustainable Housing, including affordable social housing; *(iii)* enhance Sustainability in the Building Supply Chain; *(iv)* reduce climate impact and strengthen climate resilience of the buildings and construction sector; and *(v)* promote knowledge sharing, outreach and awareness raising.

The programme involves sharing good practices, launching pilot projects, supporting projects in developing countries, emerging economies and industrialised nations alike, creating cooperation networks both regionally and internationally, and committing actors around the world to sustainable construction.

---

[77] https://www.oneplanetnetwork.org/sustainable-buildings-and-construction/about

The Sustainable Food Systems (SFS) Programme is a multi-stakeholder partnership focused on catalysing more sustainable food consumption and production patterns. The vision enables partners to collaborate on joint initiatives, which range from normative, advocacy and policy support activities, to research and development projects as well as on-the-ground implementation activities that address food systems challenges. The Programme promotes a holistic approach, taking into account the interconnections and trade-offs between all elements and actors in food systems. The planet has the capacity to provide a growing world population with enough nutritious and varied food, now and in the future. However almost 800 million people go hungry and about 2 billion are malnourished. About 30 percent of the global adult population is overweight or obese, and around 30 percent of food produced worldwide is lost or wasted. Food systems are both contributing to and affected by challenges including climate change, land degradation and biodiversity loss. They rely on a natural resource base that is becoming increasingly fragile and scarce. Unless consumption and production patterns are brought to operate within planetary boundaries, such pressures will further increase with population and economic growth. The SFS Programme has four work areas and five cross-cutting focus themes that guide the Programme towards the achievement of its goal. The four work areas are: *(i)* raising awareness on the need to adopt SCP patterns in food systems; *(ii)* Building enabling environments for sustainable food systems; *(iii)* increasing the access to and fostering the application of actionable knowledge, information and tools to mainstream SCP in food systems; *(iv)* strengthening collaboration among food system stakeholders to increase the sector's SCP performance. Each work area has a dedicated Task Force. The following are the five focus themes that are being addressed under these four work areas: *(i)* Sustainable diets; *(ii)* Sustainability along all food value chains; (iii) Reduction of food losses and waste; *(iv)* Local, national, regional multi-stakeholder platforms; *(v)* Resilient, inclusive, diverse food production systems.

And finally, the last but not least program, what we think is among the most important to pursue, since it implies a radical change in the way people think.

Sustainable Lifestyles and Education[78] programme envisions a world where sustainable lifestyles are desirable, beneficial and accessible for everyone, enabled, supported and encouraged by all sectors of society, including governments, the business sector and civil society. Today, our global footprint is about one and half time the Earth's total capacity to provide renewable and non-renewable resources to humanity. If nothing changes, in 35 years, with an increasing population that could reach 9.6 billion by 2050, we will need almost three planets to sustain our ways of living. Rethinking the ways we produce, consume and exchange has become crucial to move towards a society where we can all live well within the boundaries of our planet. As cultures and norms are core determinants of our rich and diverse lifestyles, they will need to be considered as we rethink the way societies are organised, resourced and maintained.

A "sustainable lifestyle" is a cluster of habits and patterns of behaviour embedded in a society and facilitated by institutions, norms and infrastructures that frame individual choice, in order to minimise the use of

---

[78] https://www.oneplanetnetwork.org/sustainable-lifestyles-and-education/about

natural resources and generation of wastes, while supporting fairness and prosperity for all[79].

Within this program, *The Good Life Goals*[80] highlight the vital role of individual action in achieving the ambitions of the *Sustainable Development Goals* (SDGs). People around the world are becoming increasingly familiar with it, but how many of us really know what we can do to reach them? The Good Life Goals represent an effort to answer this question and help a global audience to recognise the vital role of individual action in achieving the SDGs. It lays out 85 ways anyone can contribute towards the huge, planet-changing objectives that sit at the heart of the SDG agenda. Goals have been shaped through a multi-stakeholder collaboration[81]. The big idea is that people power matters as much as powerful people. With our voice, our actions and how we treat each other and the world around us. But we need to know what to do: the simple, collective and impactful actions people can take everywhere around the world. The Good Life Goals will give everyone a role in making tomorrow better than today. In brief, the *Good Life Goals* are:

- 85 individual actions - 5 asks for each of the 17 SDGs.
- Led by UN 10YFP & Futerra, supported by Governments of Japan and Sweden, IGES, SEI, UNESCO, UNEP and WBCSD.

---

[79] This is an updated definition based on UNEP (2010), the Taskforce on Sustainable Lifestyles (Sweden, n.d.

[80] The Good Life Goals by Futerra Sustainability Communications Ltd and 10-Year Framework of Programmes on Sustainable Lifestyles and Education Programme is licensed under CC BY-ND 4.0 — https://www.oneplanetnetwork.org/sustainable-lifestyles-and-education/good-life-goals

[81] Futerra, the 10 YFP Sustainable Lifestyles and Education program, co-led by the governments of Sweden and Japan represented by the Stockholm Environment Institute (SEI) and the Institute for Global Environmental Strategies (IGES), as well as UN Environment, UNESCO and WBCSD.

- An entry point for any government, NGO, or company in any sector, into the individual behaviours linked to activities, products and services, sustainable lifestyles, and the SDGs themselves.
- A highly engaging way of personalising and humanising the SDGs.
- Primarily designed to be used by policy-makers, business, civil society, creatives and educators who want to communicate about the SDGs.

"For the goals to be reached, everyone needs to do their part: governments, the private sector, civil society and people like you."

The Good Life Goals are a set of personal actions that people around the world can take to help support the Sustainable Development Goals (SDGs). They are lifestyle asks for individuals that are carefully aligned with the SDGs 169 targets and indicators. The SDGs have been transformational for policy-makers and business leaders in setting macro strategies towards urgent sustainability milestones that must be achieved by 2030. In parallel, a global movement for sustainable lifestyles is underway: a drive for a redefined "Good Life" involving individuals, brands, community groups, and educators. The Good Life Goals were created to bridge the gap between the Sustainable Development Goals and the sustainable lifestyles movement. Their aim is to help policy-makers, businesses, civil society groups, educators and creative professionals inspire enthusiasm, connection and action from the public in support of the SDGs. By providing personally relevant links to each SDG, the Good Life Goals send a message that we all, individually and collectively, can play an important role in the future. We all have the right, responsibility, and the opportunity to change the world for the better. Goals can help the global public to recognise the vital role of individual action in achieving the ambitions of the SDGs. They were created to have tangible impact, be relevant and accessible to the greatest number of people and be comprehensible by individuals around the world. Simple, positive, and engaging by design, the Good Life Goals detail the things that people can do, not a long list of things that they should not do.

Although ultimately written to be accessible by anyone, the most frequent users will be those who seek to engage with the public — policy makers, business, educators, creatives and community leaders. The initial objectives of the Good Life Goals were straightforward and ambitious: *(i)* increase recognition by decision-makers about how vital individual action can be in reaching the SDGs; *(ii)* provide a clear link between the SDGs and sustainable lifestyles; *(iii)* offer a tool for those seeking to engage the general public in making tomorrow better than today.

The Good Life Goals create a bridge for policy-makers between the SDGs and people's everyday lives. They do not shift the burden for action solely on individuals, nor do they simply promote new ways to consume. They highlight the important role that individuals and their choices can play. And they identify ways to meet all our needs within planetary boundaries, increasing awareness of sustainable lifestyle choices. Governments can use the Goals to raise public awareness of more sustainable choices that are available. Life Goals can be incorporated into messaging on policies and practices to engage citizens in supporting the SDGs. Such efforts can also nudge public institutions and organizations to lead by example and inspire their employees. Governments can use the Good Life Goals

114

to identify more sustainable lifestyle options and design policies and incentives to encourage the uptake of these sustainable choices. Furthermore, the Good Life Goals can help policy-makers to think beyond legislation and economic incentives, to also consider physical infrastructure, as well as the partnerships and dialogues that together can harness the power of individual decision-making and drive more sustainable lifestyles. The Good Life Goals are a tool for governments to use to help people understand the relevance of specific SDGs to their daily lives and how our personal actions interact with the world around us.

The Good Life Goals can help educators understand the relevance of specific SDGs and the crucial role education has in supporting all 17 SDGs. Innovative, learner-centred teaching and learning methods help empower learners with the knowledge, skills and values they need to address the social, environmental and economic challenges of the XXI century. The Good Life Goals provide educators with an easy tool to integrate critical issues, such as climate change, biodiversity, disaster risk reduction, and sustainable consumption and production (SCP), into their educational activities, and to motivate people to adopt more responsible lifestyles. The Good Life Goals offer different ways for learners to become empowered and take action to build a more just, peaceful, tolerant, inclusive, secure and sustainable world. The global community is now asking not only if students are in school, but what they are learning, and whether that will contribute to making the world a better place for all. The Good Life Goals offer ideas for educators to develop cross-cutting sustainability competencies in learners, as spelled out in the UNESCO publication *"Education for Sustainable Development Goals: Learning Objectives"*.

The Good Life Goals provide business with a completely new way of thinking about the SDGs and sustainability. They offer a link between what a company makes, the actions being taken to improve the sustainability of products, services and operations, and the way in which their brand exists within their customers' lives. They help business to understand how the actions and lifestyles of their customers link to the SDGs. This understanding can then help brands to more effectively engage with customers around the behaviours that are linked to their products and services in order to drive positive SDG impact. They also provide companies with a simple tool that they can use to engage staff across a wide range of internal activities, promoting enhanced awareness of the SDGs and a culture of behaviour that is more in line with their ambitions. The Good Life Goals can be leveraged to channel new product development to support more sustainable lifestyles. They help companies to understand how their products are involved in the impacts of people's lifestyles, and identify the potential for innovation opportunities. Companies can use the Good Life Goals to identify how they can offer people "better".

The Good Life Goals are an inspirational resource for creatives. They are a chance to communicate the SDGs to the world in a completely fresh way. The simple actions highlight the important role individuals can play in making the SDGs a reality, and so provide creatives with the foundation for effective communications that help people change the world for the better. The Good Life Goals are a tool for creatives to help people understand the relevance of the SDGs to their daily lives and the simple actions they can take to make help make tomorrow better than today. The Good Life Goals are simple actions that will lead to better lives and, ultimately a better world. Creatives can raise awareness of the actions and show how these will create a more desirable future for people and planet. Creatives are experts in changing people's minds and behaviours through their work. So whether it's inspiring audiences with the solutions available or nudging people to more sustainable behaviours, creatives can use the Good Life Goals to embed and normalise positive behaviours.

So, sustainable lifestyle is a state of mind, in which habits and behavioural models must be freed from the grip of unbridled consumerism, from the norms and rules imposed by the current economic paradigm.

What governs these behaviours and choices are a group of diverse and complex drivers reflecting basic needs and desires, the personal situation, socio-technical conditions and physical and natural boundaries. The drivers cover a varied range and show that lifestyle and consumption decision making is determined by many overlapping constraining or liberating factors such as cognitive abilities, psychological, social, economic, policy and institutional frameworks. Sustainable lifestyles are not then simply a matter of consumer choice, improved awareness and behavioural change, but involve the development of supporting frameworks to ensure sustainable lifestyles in the long term.   Human beings are blessed with the capacity to think ahead, explore, shape and respond to futures. This could be our greatest asset in the years ahead as we embark on an unprecedented journey to create new alternatives in the ways we live. The next decades requires nothing less than a complete transformation in our societies to stay within our ecological means and well below 2 degree temperature rise while also ensuring wellbeing for all and a just transition. For industrialised countries, ecological footprints need to be reduced a lot, as we mentioned in the Ecological Policy Handbook - Vol. I and II. The scientific evidence for this transformative change is clear including in reports by the Intergovernmental Panel on Climate Change, International Resource Panel, the latest biodiversity and ecosystem analysis, and assessment of social inequality. The magnitude of the change is also in many ways unimaginable, yet it is our imagination that is critical to enabling this transition.

Thinking and engaging with possible futures is often a missing dimension of sustainability discussions.[82] As systems expert Donella Meadows writes, *"if we don't know where we want to go, it makes little difference that we make great progress... yet vision is not only missing almost entirely from policy discussions; it is missing from our whole culture."*[83]

Futures are the visions or scenarios of possible alternative pathways for humanity. The *futuring* process brings into focus not just *what,* but *who* and *how* we engage our societal capacity for anticipation and novel ways of thinking about alternatives. There are a number of reasons why futures and the process of 'futuring' is critical at this time. Futures visioning influences our actions and reframes our perception of the present. Our images of futures influence our present actions and shape our expectations in more profound ways than we acknowledge. We can look back to the 'history of the future'[84] and note the ways in which past futures visions of, for

---

[82] https://envisioninglifestyles.org/why-futures/#_ftnref2

[83] Meadows, Donella (1994) Envisioning a Sustainable World.Published in Getting Down to Earth, Practical Applications of Ecological Economics, edited by Robert Costanza, Olman Segura and Juan Martinez-Alier. Island Press, Washington DC, 1996 http://donellameadows.org/archives/envisioning-a-sustainable-world/

[84] Hajer, M. (2017) *The Power of Imagination.* Inaugural Lecture. Utrecht University, The Netherlands.

example, car-oriented urban planning fundamentally influenced the design of cities. The visions of the *Futurama exhibit* designed by General Motors and Norman Bell Geddes and the plans by architect Le Corbusier, shape city planning from the early XX century to today, with all the resultant social, economic and ecological problems. The potential of futuring is that it can bring us together to change our stories and, in this way, to change our expectations and to catalyse the co-creation of those futures. Engaging with futures enable novel thought and get us out of stuck patterns. The challenge of transforming into sustainable societies and everyday lives requires us to think outside the box. This is not a question of tinkering with the existing economic systems but shifting our assumptions that progress is rooted in exponential resource and energy growth. When we *"look to the future with eyes tarnished by the present...everything seems huge and insurmountable"*; however, when we think together about futures, it is possible to reframe stuck debates and building shared understanding of emerging realities and common interests. *"The future is an open space still undetermined and thus less burdened by past differences, grievances and assumptions"*[85]. Futures enable us to think freely by exploring possibilities. Futures enable us to act even in times of unprecedented change and uncertainty. The nature of the current challenge requires high-quality futures thinking. As sociologist Jens Beckert presents the concept of *"fictional expectations"* and notes their key role in enabling action even in times of great uncertainty by creating collectively held images of futures and how they might

[85] Boyer, N. and V. Timmer (2012) *Envisioning Sustainable Futures*. In The State of the World 2012: Moving Towards Sustainable Prosperity. Worldwatch Institute, Washington DC, USA.

evolve.[86] In his analysis of capitalist economies, he explores how these predictions become self-fulfilling prophecies by generating the expectation of how markets may evolve even in times of great risks and opportunity. The futuring process allows us to rehearse possible options and trajectories. Futures thinking is typically linked with forecasting and prediction; however, there are significant limitations in simply projecting current trends into the future. Futures techniques such as scenario-planning enables a richer approach that explores diverse story-lines of what 'could happen' under different conditions. Backcasting techniques explores desirable end-states and remain flexible about pathways to get there. In this way, futuring allows us to rehearse these different trajectories and dialogue about their possible implications. Futuring processes can have democratic value. Futuring is not only the domain of experts but can also be deeply participative in their design. John Robinson notes that sustainability is actually 'procedural sustainability' in which *"sustainability can use usefully thought of...as the emergent property of a conversation about desired futures that is informed by some understanding of the ecological, social and economic consequences of different courses of action."*[87] The opportunity lies in generating participative visioning processes result in concrete forms of anticipation accessible to all publics and likely to enable both formal deliberative processes and informal social conversations on the future at societal level as well as empowerment of citizens in education for responsible living and democracy. The challenge is that our mainstream depictions of the future are fre-

---

[86] Beckert, J. (2016) Imagined Futures – Fictional Expectations and Capitalist Dynamics. Cambridge, Mass.

[87] Robinson, J. and R. J. Cole (2015) Theoretical underpinnings of regenerative sustainability. *Building Research & Information*. 43 (2): 133-143.

quently dark and dystopian — futures where the deprivation, oppression and terror are the primary focus. There is growing evidence that fear-based messaging — emphasising a problem or threat — is not always effective in stimulating behaviour change and can actually lead to defensive avoidance[88] and psychological distancing[89]. On the other hand, there are indications that positive approaches lead to greater engagement and excitement and more successful and longer lasting change[90]. Aspirational stories change the lens through which we view reality, spark imagination, and appeal to the non-rational and emotional aspects of human decision-making.[91] This does not mean that the problem and challenges are ignored. Aspirational approaches only resonate when solutions and vision are presented as a response to the magnitude of the challenge they aim to address and overcome. There is power in clearly articulating what gets better with sustainable, low-carbon futures[92] and in

[88] Witte, K., & Allen, M. (2000). A meta-analysis of fear appeals: Implications for effective public health campaigns. Health Education & Behaviour, 27, 591-615; Van't Riet, Jonathan and Robert A.C. Ruiner (2011) Defensive reactions to health-promoting information: an overview and implications for future research. Health Psychology Review. Vol 7, Aug.

[89] Pike, Cara, Sutton Eaves, Meredith Herr, Amy Huva, David Minkow (2015) The Preparation Frame: A Guide to Building Understanding of Climate Impacts and Engagement in Solutions. Climate Access. March. http://www.climateaccess.org/resource/preparation-frame

[90] Coghlan, Anne T., Hallie Preskill, Tessie Tzavaras Catsambas (2003) An Overview of Appreciative Inquiry in Evaluation. New Direction for Evaluation. No. 100, Winter; Whitney, Diana and Amanda Trosten-Bloom (2010) The Power of Appreciative Inquiry: A Practical Guide to Positive Change. Berrett-Koehler Publishers San Francisco, CA.

[91] Korten, David (2015) Change the Story, Change the Future: A Living Economy for a Living Earth. Berrett-Koehler Publishers Inc.

[92] Cara Pike, Sutton Eaves, Meredith Herr, Amy Huva, David Minkow (2015) The Preparation Frame: A Guide to Building Understanding of Climate Impacts and Engagement in Solutions. Climate Access. March. http://www.climateaccess.org/resource/preparation-frame

describing futures in a way that presents a compelling vision of less material ways of meeting needs and aspirations.[93]

"Where in the past, we focused on wealth, growth and efficiency; the future will need to be about well-being, quality, and sufficiency. This includes living within limits; shaping a sustainable society (not just a sustainable consumer); addressing the public as citizens, not consumers; addressing production and consumption; and creating the systems that lead to sustainable behaviour . . . yet not everything is about reduction — there are some things that are not near peak or have no limited supply: community, personal autonomy, satisfaction from hones work well done, intergenerational solidarity, cooperation, leisure time, happiness, ingenuity, artistry and beauty"[94].

This is a critical time to engage in collectively imagining compelling futures in order to manage critical transitions ahead. While we imagine these futures, we also need to be focused on how these visions reflect the core sustainability foundations including our ecological impact and economic and social imperatives. The majority of sustainable futures can be broadly categorised in terms of four tendencies: *(i) smart green techno-living*, with an emphasis on technology, efficiency, gadgets that support daily living, and high consumption lifestyles; *(ii) sustainable urban and rural design*, literal greening through trees, plants and living walls; a focus on green buildings, mobility, energy systems, urban farming and industry clusters; *(iii) eco-communities*, alternative ways of living including alternative housing such as co-housing and earth houses, walking with some bikes and transit, gardens, social connectivity and sometimes diversity; *(iv) living green*

---

[93] Jackson, Tim. 2009. Prosperity without Growth—Economics for a Finite Planet. London: Earthscan.

[94] Fedrigo, Doreen and Arnold Tukker (2009) "Blueprint for European Sustainable Consumption and Production: Finding the path of transition to a sustainable society", European Environmental Bureau, May. P. 9.

expos and grade show, a focus on the gadgets and products of daily living rather than practices, emphasises the separate domain of consumer goods; and focused on individuals rather than collective solutions. There is value in these tendencies. Each of these approaches have key elements of sustainable futures. Sustainable futures will certainly require the smart technology systems of *Smart Cities*, the focus on greening and mobility and housing of *Smart Urban Planning*, the innovative alternative ways of living of eco-communities, and cutting edge sustainable goods and market solutions that are displayed at trade shows. The transformation to sustainable living will need much more than this and sustainable living futures will need to stretch far beyond these current depictions.

Terms like "quality of life" and "sustainable lifestyles" regularly appear in the media, illustrating that people are already weaving sustainability into their daily decision-making. Carbon foot-printing, food waste reduction campaigns, urban gardening, vehicle sharing models, and surveys to understand the values and motivations of youth are all ways that are helping people to live more sustainable lifestyles. Yet these actions, in general, are piecemeal. They are not yet framed within a holistic vision of what constitutes a sustainable lifestyle. However, it looks like there is a solid foundation in place, and it is now time to develop a more structured life-cycle, and evidence based understanding of sustainable lifestyles to facilitate global dialogue and measure progress. This will enable us to focus on the 'hotspots' on where critical action can be taken. For individuals, it means understanding the impacts of their daily decisions and embracing more sustainable lifestyles. For governments, it implies setting a conducive regulatory context, facilitating and inspiring better citizen decision making, creating market demand through sustainable public procurement, and supporting research, development, and innovation. For the private sector, it implies integrating sustainability into core business strategies to develop innovative ways to meet the needs of people while reducing the pressure on the world's dwindling resources. This includes communicating about product sustainability performance to enhance informed decision-making.

Key lifestyles domains and the environment

Given that consumption is heavily embedded in lifestyles, researchers have been able to identify key domains where consumption and lifestyles have the highest environmental impacts by combining an understanding of consumption patterns, life-cycle analysis and sustainability indicators for carbon, material, and ecological footprinting. For example, the *International Resource Panel* (IRP) produced a synthesis report[95] with a global assessment of final consumption categories and product groups that have the highest environmental impacts across their life cycle. The top three impact categories are *food and agriculture, housing and building construction, mobility and transportation.* Other studies which focus on national[96] and regional[97] levels draw similar conclusions. In addition they all highlight the large footprints of consumer products and services, including those related to tourism and entertainment. There are, however, limitations to relying only on footprint calculations for policy design and action. These include: problems with data, the number of individual product life cycles that would need to be analysed to provide a

---

[95] Hertwich, van der Voet, & Tukker, 2010

[96] e.g., Lettenmeier, Liedtke, & Rohn, 2014; Michaelis & Lorek, 2004

[97] Backhaus, Breukers, Mont, Paukovic, & Mourik, n.d.; EEA, 2012; OECD, 2002

comprehensive impact assessment, and the fact that most assessments carried out reflect consumption in industrialised countries and regions. For a broader picture of lifestyle impacts, quantitative methods (such as footprint analysis) need to be complemented with normative, qualitative assessments, especially in the case of emerging economies. This report highlights that sustainable lifestyles imply more than material consumption alone. Beyond just environmental impacts, the social impacts of lifestyles and consumption can be equally or even more problematic. For the purposes of developing research-based, practical strategies and responsive interventions towards sustainable lifestyles, the key domains of final consumption highlighted in this report are: food, housing, mobility, consumer goods and leisure. Water, energy, and waste are not addressed in isolation but as cross-cutting elements that affect and are affected by almost every lifestyle domain. The emerging practice of hotspot analysis to highlight where action could be taken along supply chains and potential impacts, though not covered in this report, is instrumental in the next steps[98]. Similarly, understanding the political economy around consumption and the power dynamics in the supply chain would further highlight which stakeholder(s) has the most potential for sustainability

Food

What we eat and drink — how it is produced, processed and provided — and how we dispose of it all have impacts on the environment and society. People make decisions related to food based on both objective and subjective factors, including cost, freshness, health impacts, presentation (e.g., packaging), place of origin, convenience, taste and culture. At the use phase in the food system,

---

[98] Barthel et al., 2014; Tukker et al., 2006; UNEP, 2010b

some factors that have impacts on the environment include outlet of purchase, storage period and facilities, preparation process and consumption. Apart from environmental impacts, concerns around lifestyles and food include health, obesity, an increasing number and intensity of allergies and social impacts of agricultural practices. Globally, almost a third of food harvested is wasted or lost; contributing to this are changing dietary trends, particularly in urban environments which increasingly favour more resource intensive foods (GHG producing) such as processed foods and meats. This occurs in a global context where 1 in 9 people are hungry and 2 in 10 are obese. There is clearly potential to shift to more sustainable patterns. Cities can encourage more sustainable diets that ensure adequate nutrition while reducing environmental footprint, raising awareness, and changing behaviour around food waste. Enacting policies in planning, housing and transportation can also support more sustainable low carbon food systems and encourage more sustainable local food production such as backyard and community gardens.

Housing

How we live, where we live, what is used to build, heat and cool our living spaces and what we install in our houses, have social and environmental impacts. The building sector contributes up to 30 per cent of global annual greenhouse gas emissions and uses up to 40% of all energy[99]. In order to address this, we need innovative solutions on what future buildings and cities will look like. Building construction requires resources such as sand, wood and metals. Many of the materials require preprocessing and some of them are sourced through mining. The mining process alone causes biodiver-

---

[99] UNEP, 2009

sity loss, deforestation, emissions of GHGs and use of hazardous chemicals. People make decisions related to housing based on both objective and subjective factors, including cost and size of the building, building characteristics, aesthetics, the neighbourhood and available amenities. While living in houses we use energy and water, and dispose of waste: important energy considerations include efficiency insulation and heating and/or cooling. The way neighbourhoods are built affects many aspects of society, including the rate of crime, commuting distances, the opportunities for neighbours to create strong ties and form vibrant communities and the general well-being of inhabitants. Finally, at the end of a house's life cycle, the building needs to be demolished, requiring energy and producing waste.

Mobility

What forms of transport we choose, how often we travel, and the distance travelled as well as the supporting systems and infrastructure have impacts on society and the environment. The transport sector is responsible for 13 per cent of greenhouse gas emissions and 23% of CO_2 emissions from global energy consumption[100]. Citizens make mobility decisions based on cost, choice of transportation mode, congestion, convenience, time efficiency, connectedness and environmental impacts. Mode of transportation is particularly significant — flying tends to have the highest environmental impact, followed by private car use. Other factors, such as distance covered, number of people in the vehicle per use, technology efficiency and type of fuel used, also contribute substantially. Awareness of climate change has led to increased understanding of mobility impacts on society and the environment. However, more can be done to under-

---

[100] GEF-STAP, 2010

stand how people's mobility needs can be addressed in a more sustainable way. Part of this involves questioning the need for mobility in given situations, and in making living choices that require less transportation (e.g., residential housing configurations that require less commuting and less travel for shopping and entertainment). For example, policy responses can include combinations of measures that discourage unnecessary transportation, adopt more sustainable modes of transport, and improve existing systems of transport.

Consumer goods

The products we buy, the type and quantity of materials that are used in producing them, how we use them and how often we replace them, have impacts on society and the environment. Examples include electric and electronic appliances, clothing and shoes, cosmetics and personal care, jewellery, furniture and paper products. Products which tend to have the highest impacts are those produced using mined materials and fossil fuels. Consumer goods are important because of their daily use and their role in defining our image and habits. The expanding role in modern lifestyles of electric and electronic products (e.g., mobile phones and other information communication products) means related environmental impacts are increasing, through the growth of electronic waste, pollution and mining of rare earth metals. However, they also have a strong potential to unleash sustainability efforts because of their role in disseminating information and enhancing knowledge/experience exchange and their global reach. The fashion and textile industries have evolved to embrace a fast-fashion phenomenon, characterised by tens of "micro-seasons" and some popular brands introducing hundreds of new styles a week. For some middle class consumers, clothes are rarely, if ever, worn and move quickly to landfills. These consumption patterns have huge implications for resource scarcity and pollution, with impacts that vary according to fabrics, dyes, chemicals, transportation, and packaging method used. Clothes help us to define who we are and what we stand for, and are connected to our daily lives on a very personal level. With women spending tens to hundreds of hours shopping for clothing every year, fashion has the unique ability to be a highly visible engine for change and even a medium for consumer education.

Leisure

How we spend leisure time, our choice of tourism destinations and activities, and the facilities we use have significant contributions to the environment and society. Leisure embodies a wide variety of activities — from meditation and reading to flying and watching television; or swimming, golfing, weekend trips, and owning second homes. Each reflects different levels of materialism and social interaction. Staying at and using the services of a five-star hotel, for example, has a higher impact than staying in a three-star hotel. Entertainment activities increasingly involve electronic equipment and information communication technologies; this has led to higher levels of individualism while at the same time increasing energy use and electronic waste production. Tourism products and services, if not well managed, can contribute to biodiversity loss, stress on key resources, land fragmentation, social disruption, and loss of cultural heritage. On the other hand, volunteering for social causes, meditation, and engagement in handicrafts, when sustainably managed, have been shown to contribute to a better sense of wellbeing and social cohesion.

In general, from research on sustainable lifestyles we can indicate some key messages:

1. There is no universal sustainable lifestyle. What is sustainable in one locality may not be sustainable in another.

2. Lifestyles occur within — and are enabled and constrained by — social norms and the physical environment. It is important to differentiate between the factors that can be addressed at the individual or the household level, and those that are beyond individual control[101].

3. Lifestyles are not static. They change with society's dynamism. People's visions and aspirations in life change as their personal situation evolves, as society evolves and as knowledge, norms, and technology change[102]. These offer opportunities for shaping the future.

4. Needs and desires are influenced by time and society. As society evolves, or becomes more complex and/or affluent, what constitute basic social needs evolve. For example, a mobile phone was perceived as a luxury two decades ago, now it is a perceived need for most adults in industrialised cities, yet it remains a luxury in some parts of the developing world.

5. Beyond enabling basic necessities and needs to operate with dignity within a society, increases in income not directly translate into happiness. People's expressions of happiness correlate with the level of trust, social ties, education, health and meaningful employment[103]. There is little evidence, especially in industriali-

---

[101] Akenji, 2014

[102] UNEP, 2011

[103] Easterlin, 2003

sed nations, to support the assumption that economic growth through gross domestic product translates to increase in wellbeing[104].

6. Efforts must be made to address the extremes of poverty and wealth in society in order to ensure sustainable lifestyles. Manifestations of social tension get stronger as the disparity of economic conditions between the social classes get wider[105].

7. The environmental impacts of lifestyles are not intentional but rather a consequence of people aspiring to fulfil needs and desires, as well as to function in society. It is important to examine how society is organised to provide for the wellbeing of citizens[106].

8. Most environmental impacts of lifestyles can be addressed by targeting the following key domains of final consumption: food, mobility, housing, consumer goods, and leisure. This cannot be done piecemeal and must address the underlying value systems (including what contributes to well-being) and review the choice of architecture and infrastructure that support lifestyles.

9. Knowledge or awareness of sustainable consumption and lifestyle options does not usually lead to intended actions. This knowledge-action or intention-behaviour gap suggests that awareness is easily subordinated by lack of access or lock-in to available options.

10. Top-down approaches to changing lifestyles will only succeed with participation of civil society. Bottom-up approaches, inclu-

---

[104] Jackson, 2009

[105] Death, 2014; Hilton, 2007

[106] Shove, 2006; Spaargaren, 2004

ding social innovations, social movements, and grassroots experiments, are pivotal in opening up new avenues and fostering acceptability of sustainable solutions[107].

Influencing factors of consumption and lifestyles[108].

What works or does not work is still subject to experiment and debate. There is consensus that, to have more effective sustainable lifestyles policies and practices, it is critical to get context-specific to understand why people consume and what shapes their related behaviours. This context-specific understanding can be derived through three interlinked underlying lifestyle factors.

Motivations refer to the immediate personal and social reasons and justifications that compel people and society to take certain actions or make certain decisions — e.g. the desire to spend time with friends and family, or the seductive presentation of a product. Why do people consume? Studies and empirical evidence suggest that people do not consume with the intention to harm the environment. Resulting environmental impacts are an unintended consequence of the pursuit of well-being. Viable approaches to changing lifestyles need to address underlying reasons and motivations for particular consumption patterns. Among other reasons, people consume: to meet basic needs e.g. nutrition and subsistence, health, housing, mobility; to fulfil social functions/expectations e.g. convenience, connectedness, maintaining relationships, traditions; to satisfy personal desires, preferences and tastes e.g. leisure, food preferences, consumer goods (electronics or cars); due to the influence of

---

[107] Heiskanen, Lovio, & Jalas, 2011

[108] There is vast literature addressing lifestyles and consumption and sustainability. Akenji, 2014; T Jackson, 2005; Mont & Power, 2013; OECD, 2002; Tukker, Cohen, Hubacek, & Mont, 2010; Vergragt, Akenji, & Dewick, 2014

advertising/marketing e.g. creation of new product markets such as pet food and cosmetics, planned obsolescence, or enhanced functionality such as mobile phones that do more than make calls; because they have no choice e.g. lock-in design of mobility infrastructure favours private car use or urban zoning laws and administrative procedures make urban agriculture difficult.

Drivers refer to circumstances that support motivation, normalising it, or making it practicable — e.g. cultural norms or media marketing. Lifestyles and consumption are governed by a set of complex and dynamic drivers, which reflect the personal situation (income, identity, individual taste, and values) and external socio-technical and economic conditions (culture, social context, peer pressures, etc.). There are also physical or natural boundaries which allow or constrain lifestyle options. Studies on consumer decision-making in several fields show that cognitive abilities, psychological, social, economic, and policy and institutional frameworks all come into play, highlighting that driving factors behind lifestyles are inter-linked, and sometimes contradictory.

Let's try to imagine a ring containing in the centre different life-styles surrounded by layers of influence on needs and wants. In the centre we have needs and desires of people. Around needs and desires we have our personal situation and socio-technical condition as: *(i) income level.* This is one of the strongest lifestyle indicators and drivers of consumption. More disposable income means greater affordability of goods and services and easier access to more credit, that can further consumerism[109]. In addition, there is compounded social pressure to maintain lifestyle levels once adopted; *(ii) values*

---

[109] Girod & De Haan, 2010; Tukker et al., 2010

are powerful determinants of attitudes and actions.[110] Many consider them the foundation of lifestyle decisions as people tend to consume to fulfil value-laden objectives. Values can be at the personal or broader cultural or ethical levels[111]; *(iii)* people's *abilities* are influenced by many things (age, geography, climatic conditions, which in turn affect lifestyle decisions)[112]. For example, cognitive and physical abilities influence health, fitness, capacity and related decisions like which health procedures to undertake or which sport activities to participate in; *(iv) Awareness* is important to enable people's search for suitable lifestyle alternatives. Awareness of consumption impacts, at the individual and collective levels, can shape choices and can have a multiplier effect: groups can lead by example and an individual can influence family, friends or communities in contact. Awareness on its own is not enough — it must be accompanied and channelled (e.g., by policy, incentives, etc.) towards actionable outcomes. While awareness can change behaviour, sometimes practice or experimentation alters awareness[113]. Hence, awareness of environmental impacts is not a major determining factor in lifestyle choices, which is an assumption made by many awareness-raising campaigns[114]; *(v)* the availability (or the lack) of *knowledge* and information on products, services, and alternative options can often encourage or hinder lifestyles choices. Knowledge is influenced by formal and informal education, employment (type of job), and ex-

---

[110] Brodhag, 2010

[111] Mont & Power, 2013

[112] OECD 2002

[113] Guagnano, Stern, & Dietz, 1995

[114] Akenji, 2014

posure to informal information sharing such as media, family and friends[115]; *(vi) social norms and peers*: Our lifestyles are heavily influenced by those around us: family background, social circles, colleague expectations, professional decorum and social practices, etc. As social beings, humans have a need to identify with groups and there is peer pressure to fit in and engage in similar activities, rituals, conspicuous consumption, etc. There is also a tendency within the emerging culture of mass customisation, for people to differentiate themselves (to a limited degree) to express uniqueness[116] or a status level within a social group hierarchy. Social and cultural institutions are custodians of culture and adherents to principles that propagate value systems, and hence are important in shaping values, social norms and lifestyle choices; *(vii) the media* with its far reach into our lives is one of the strongest influences on values, social norms and lifestyles, spreading and accelerating the social norms of consumerism. Advertising and marketing strategies often help create new (sometimes false) 'needs' and trends, encouraging consumers to replace still-functioning products for newer ones[117]. With increasing exposure to different media channels, including social media, facilitated by technology, the role of media to shape consumer preferences is steadily getting even stronger; *(viii) market prices* determine who can afford market options. Thus pricing of luxury goods or sustainable products predetermines who can access them. When more sustainable products or services are priced higher than the less su-

---

[115] Barth et al., 2012; UNDESA, 2010

[116] Baudrillard, 1998

[117] Cooper, 2004

stainable alternatives, the sustainable option is less competitive.[118]
As disposable income increases, people are less susceptible to price
variations (expensive or luxury goods can become relatively more
affordable). Hence the perception that higher priced organic or fair-
trade products are fashionable items for the wealthy; *(ix) technology*
can change ways of doing things[119] — shopping by internet and e-
commerce are key examples. Characteristics such as complexity,
resource efficiency, indigenousness, and affordability influence the
uptake and use of technologies. As they get into wider use, they often generate new eco-systems, such as supporting products, new systems of provisions, infrastructures, social practices and even subcultures. For example, mobile phones now often come with accesso-
ries such as casings and purchasable apps and new communities or
subcultures around the apps. Technology can raise standards of living, e.g., through electricity, agriculture, and communication, but
are also known to be coupled with unsustainable production and
consumption patterns, which result in higher overall consumption
of natural resources, goods, and services; *(x) infrastructure:* this refers
to the hardware such as buildings, provision systems for water and
sewage, electricity, waste management, telecommunications networks, and public transportation networks. They tend to have long
lifespans and their designs lock people into specific use patterns,
hence getting their design right from the start is important[120]; *(xi)*
policies and institutional frameworks: these have a powerful influence on
all stakeholders and lifestyle directions. Hard (e.g., penalties and

---

[118] Alcott, 2008; Godfray et al., 2010

[119] Christensen et al., 2007; Shove, 2004

[120] Kivimaa & Mickwitz, 2011; Sahakian & Steinberger, 2011

subsidies) and soft (e.g., nudging and voluntary standards) policy instruments can shift the entire consumption architecture by changing available market options, editing out less sustainable options, encouraging more sustainable alternatives, and creating platforms for innovation by both businesses and society. It has been argued that, "the most significant agency is usually found in addressing the wider contextual issues, for instance by changing the law or by amending the public procurement process for major projects such that sustainable development issues may more reliably be incorporated in the design"[121].

Determinants are super-factors that decide on the possibility of lifestyle or consumer action. These should be the focus of policies, institutional frameworks, programmes and infrastructure when influencing lifestyle design. While motivations and driving factors explain the need or desire for a particular lifestyle or consumption practice, they translate into action only when certain determinants are in place. Determinants are metafactors that establish whether or not a lifestyle is practiced and/or sustainable. Based on their characteristics, determinants can be grouped as: *(i) attitudes* are a cluster of factors that contribute to a person's value orientation and their likelihood to consume. They determine preferences and choices — e.g. people who are health-conscious and eat less meat or are vegetarian tend to express pro-environment or religious attitudes. They include cultural ethics, social norms, professional and peer principles, media messages, and awareness. They create an 'appetite' tailored towards a particular direction. Attitudes can refer to individual orientation as well as collective social values and are heavily influenced social practices and movements; *(ii) facilitators/access*

---

[121] Ballard, 2005; Pg 143

means that belonging to a community network can facilitate access to certain local goods or services. In the same vein, a government policy can facilitate development of more competitive renewable energy options. Facilitators are a set of factors that contribute to the possibility for certain behavioural patterns or a lifestyle to actualise. Having a propensity to lead a consumerist lifestyle is not enough; one must have access to the consumer goods and services, social networks, etc., that make up that lifestyle. Access reflects 'agency,' or the ability to take personally meaningful actions. This manifests through the availability of options or choices that allow for tailored responses. Purchasing power (e.g., through income), availability of time, social networks, and cognitive and physical abilities can all contribute towards access; *(iii) infrastructure* refers to socio-ecological interfaces that support consumption activities. They include the physical infrastructure (for housing, mobility, and leisure) and the design of systems of provision (e.g., pricing and capacities of utilities like water and energy). Infrastructure around housing and transportation, for example, would need to be accessible, safe, dependable, etc., and because it lasts a long time and tends to lock users into particular behaviour pattern patterns throughout their operational lifespan, need to be highly sustainable.

Overall, the three sets of influencing factors can be seen in ascendancy, based on the impact they have on consumption and lifestyles, starting from having the motivation, to the drivers and to the presence of determinants.

Encouraging bottom-up action: the REDuse framework

Understanding the factors that influence lifestyles allows for more strategic design of targeted sustainability interventions. Though lifestyles are primarily manifested in individual actions, support is required of all stakeholders including governments, businesses and institutions. There are two approaches to assess and design sustainable lifestyle policies and actions. The *Refuse, Effuse and Diffuse* (REDuse) framework, supports bottom-up approaches, encourages programmes and actions that directly empower individuals and households in their daily lives (and, indirectly, communities), enabling them to understand, create and/or choose the more sustainable lifestyle options. The *Attitude-Facilitator-Infrastructure* (AFI) framework is a top-down approach to support government policy, business models, institutional arrangements, and actions that set the conditions necessary for sustainable lifestyles to thrive. Sustainable lifestyle interventions that target individuals or the grassroots level have three basic components. The first involves targeting change of individual behaviour that perpetuates negative impacts on the environment or society. This is referred to as the Refuse component. Examples could include reduction of food waste or buying over-packaged products. The Effuse component, seeks to encourage behaviours that have minimal and/or positive impacts. Using a bicycle instead of a private car or composting of organic waste are examples. While the Refuse component discourages harmful choices, Effuse encourages positive behavioural aspects. The third component Diffuse goes beyond the individual and seeks multiplier effects through engaging communities in collective sustainable behaviour. An example is in sharing or collaborative consumption — such as community gardens or farms and car-pooling. Together Refuse, Effuse and Diffuse form components of the REDuse framework. Centred on everyday sustainability actions, REDuse brings together a complementary set of practices that gradually expand from those taken by individuals to engagement on a community level.

The REDuse framework can be used to develop complementary actions in different areas and at different levels, for example, by national and municipal governments, for campaign organisation, by businesses and for/by citizens. A city could develop action plans and activities for itself (through multi-stakeholder consultations and workshops). The REDuse framework supports individual, household, and community actions and is good for campaigns and communication, but it alone cannot deliver sustainable lifestyles across society. More is needed to address lifestyle determinants, including the social and physical conditions beyond individual control. A broader strategy is also needed to engage business and institutional interests and government policy and planning to assure preconditions for sustainable lifestyles. The AttitudesFacilitators-Infrastructure framework addresses these dimensions.

The Attitude Facilitators Infrastructure (AFI) framework.

It draws from the three lifestyle determinants and describes the elements needed to design a sustainable lifestyles policy package at a systems level: pro-sustainability stakeholder attitudes, facilitators or access to sustainable options, and the supporting infrastructure. It focuses on changing the context that shapes lifestyles, addressing the macro-factors beyond an individual's control. It is aimed at governments and other major stakeholders, to aid in the design of policies, programmes, and actions that "edit out" unsustainable options and to make sustainable lifestyles the default option[122].

The *right attitude* refers to a set of positive values that lead to a predisposition to act sustainably. Attitudes include those of individuals and other influential stakeholders — businesses, policymakers, legal practitioners, farmers, designers, community leaders, politi-

---

[122] Akenji, 2014

cians, teachers, and so on. Attitudes are shaped by knowledge and value orientation. Stakeholders should be instilled with attitudes that demonstrate a comprehensive understanding of the sustainability agenda and the need for system change from top to bottom. Governments are uniquely well-placed to set the conditions for sustainable lifestyles to flourish, addressing how policies can promote or contradict this objective, and how competing influences on policy design reflect different interests as well as the power dynamics of winners and losers in society. Businesses need to understand and communicate the needs their products or services serve, what lifestyles they promote, and the impact these products and services have on the environment and society. Communities need to understand what social norms they promote and how they influence citizen lifestyle decisions. Individuals and households need to understand the impacts of their choices, the potential available alternatives, and recognise that solutions — while difficult at the individual level — may contribute to the sustainability of the larger society and environment. Optimally, all stakeholders should understand the importance of sustainable lifestyles and the attitudes needed to make them a reality. Beyond technical and marginal adjustments in production-consumption systems, realigned attitudes towards more sustainable lifestyles requires that citizens, businesses, and policy-makers learn to imagine a world in which some people consume less, while those who still need to meet basic needs consume in a way that is different from contemporary materialism. Civil society organizations play a key role here to create awareness, and create platforms for association, and to ensure acceptance and continual generation of new solutions.

Facilitators create or provide access to an enabling environment for sustainable lifestyles. They are a set of mechanisms, such as regulation, legal platforms, administrative process, market facilities, or institutional arrangements that provide incentives or constraints for sustainable options. Institutions like religions, associations, or schools are custodians of our cultures. They validate norms and shape ways of thinking and acting. Thus, should they espouse pro-sustainability principles, policies, and practices, they can inform and encourage sustainable lifestyles. Price is a good facilitator — affordable sustainable options are more attractive to choose. Product standards and consumer information are examples of facilitators that, if properly administered, could help prevent 'greenwashing' — which has a reverse effect on consumer trust — and enable citizens to make well-informed and more responsible choices. Similarly, administrative procedures can be a deterrent or facilitator of change — making access easy to more sustainable food options would encourage its consumption by default. For example, instead of requiring organic produce vendors to jump through administrative hoops (for eco-labels) to market their produce as exceptions, the logic could be reversed: the less sustainable options should get a non eco-label and shelf-placement restrictions, while the more sustainable option gets shelving priority and easier access to the market. Laws and government policy are some of the strongest facilitators. In the same way that subsidising fossil fuels provides a perverse incentive for private car use, removing the subsidy and charging a carbon tax for car use could generate revenue for, and provide incentives for public transportation development and use. Measures seeking to engender sustainable lifestyles should target the specific patterns that need to be changed. What works for one lifestyle

142

group might not affect another, or might affect another in a counterproductive manner. As an example, raising the prices of utilities to reduce wasteful water consumption might disproportionately hit those who cannot afford to pay and the price difference might not be high enough to dissuade the overconsumption behaviour. Any policy package must therefore address utility prices, design of the provision system and the factors that influence use patterns of different peoples.

Infrastructure includes the products and services being consumed, the social environment and physical infrastructure that foster sustainable behaviours. Even if all citizens sought to live sustainability, this would not be realised without more sustainable product options that are comparably safe, of similar quality, healthy, accessible and reasonably priced. Given its influence on behaviour, and how long it tends to last, the design of infrastructure for domains such as food systems, housing, mobility and leisure, is critical to sustainability. The design of utility systems, for example, has implications for resource consumption at home. Toilet tanks, for example, generally flush more water than needed for each use, and buildings with automated motion-detecting switches consume comparatively less energy. In addition to characteristics of individual units, the configuration of infrastructure systems influence sustainability of its use. Zoning laws that promote the development of residential areas far from work places and shopping areas encourage frequent travel, which can be particularly unsustainable if there is little or no accessible public transportation. Businesses and investors are instrumental in ensuring that infrastructure promotes sustainable lifestyles. In addition to sustainability standards by governments, public private

143

partnerships for priority sectors can communicate on and ensure availability of infrastructure.

The *Attitudes-Facilitator-Infrastructure framework* helps government work helps government understand and plan how best to support more sustainable lifestyles. To empower REDuse (*citizen actions to Refuse, Effuse and Diffuse*), user platforms for innovation and to create socio-ecological interfaces that promote concerted societal engagement for sustainable lifestyles, are needed. City governments can set up the following facilitating efforts or 'facilitators' to create more enabling environments:

- Citizen panels on innovation for sustainable lifestyles. A broad-based platform of citizens, consumer and lifestyles organisations, public institutions, schools, etc. Such panels would co-create a shared vision of lifestyles in the city, be engaged in problem diagnosis, deliberative policy formation, proposing solutions, and facilitation of buy-in from citizens and stakeholders;

- City Index for sustainable urban living. An alternative or at least a complement to the economic growth indicators. Cities could develop a new indicator that consolidates environmental, social and economic elements into a common frame work, which inspires and reports on how urban planning, infrastructure development, policies and programs support healthy, safe, accessible and sustainable lifestyles;

- Ombudsman for sustainable lifestyles. A body that would support sustainability considerations in public decisions and infrastructure, support development and use of the city index for sustainable living and promote initiatives by the citizen panel. Such an institution could work with banks and local organisations and communities to intervene against predatory financial or lending schemes likely to cause personal, social and ecological distress;

- Business hub for sustainable lifestyles. A hub that could promote new models such as servicing, social enterprises, co-ops, repair and second-hand shops, etc.; address advertising and marketing in the city, such as limiting ads targeted at children and schools; commercial or ad-free zones/cities; using fact- and science-based claims; reduction of emotional language, etc.

The *"A framework for shaping sustainable lifestyles, determinants and strategies report"*[123] synthesises recent science-based narratives on what determines lifestyles and how they could be better shaped to respond to sustainability challenges. Lifestyles influence and are influenced by social norms and the physical environment, acting as either constraints or enablers to the many decisions citizens make every day. The so-called knowledge-action or intention-behaviour gap suggests that awareness cannot easily be acted upon if there is a lack of sustainable options and access to them, or a lock-in to unsustainable options. Raising awareness is only a part of what needs to be done. In designing actions to promote lifestyles, it is important to differentiate the factors that can be addressed at the individual or the household level, and those that are beyond individual control. Solutions need to target individuals and households as well as the stakeholder groups (communities, businesses, institutions, and governments) that shape the context of consumption and lifestyles. The *Attitudes-Facilitators-Infrastructure framework* offers a top-down policy-guiding approach to create an enabling environment within which sustainable lifestyles can flourish. To complement this, the *Refuse*

---

[123] United Nations Environment Programme, 2016

145

Effuse Diffuse framework supports bottom-up engagement by individuals, households and communities to seek personally meaningful solutions and engage in grassroots experiments and social innovations. One crucial step to support sustainable lifestyles requires understanding the patterns of different types of lifestyles — known as lifestyle segmentation. Each lifestyle segment has and manifests distinct values, preferences, and practices in areas such as fashion, use of language, and leisure activities. Earlier approaches to lifestyle segmentation have focused mostly on wealth, income and profession to establish different social classes. However, our ever-changing society and recent environmental challenges (i.e., climate change), underscore how classical approaches alone are not sufficient to address sustainable lifestyles. There is a need for countries, regions and cities to conduct social-ecological segmentation to design targeted interventions solutions and better responses. Finally, sustainable lifestyles do not always have to involve new ways of doing things, or be related to consumption. Traditional practices, old technologies, and communities living fulfilling lives without being heavily consumptive can be instructive towards formulating large-scale solutions.

Chapter VII
Trusting people

Broad public participation and the notion of trust.

Since social trust is a very abstract and rather ambiguous notion, several theoretical frameworks and approaches conceptualising trust in sociological literature have been developed. Among the many typologies[124] used, we can identify a key distinction regarding the notions of trust: *trust as a moral or emotional trait* deriving from a very early socialisation phase vs. *trust as a rational response* that is learned with a set of normative rules. In addition, we will discuss a third perception: *trust as a cultural rule.*

According to the first approach, trust is a disposition that hinges on emotions, self-perceptions, as well as ideals and values pursued in social relations[125]; and it is as much an interpretation of oneself as of the other[126]. This approach considers trust as an inevitable and natural feature of every human, which evolves from interactions with and interdependence among other humans in the society. We create ourselves as human beings through communication and interaction, and trust is a vital prerequisite of being social[127]. This perception of trust is essential to Durkheim's notion of *"collective conscience"* — a specific type of common moral beliefs that allow for

---

[124] In the literature trust is categorised as a moral trait (Uslaner, 2002), emotion (Rotter, 1971), a relationship (Hardin, 2006), an action (Sztompka, 1999), one of the elements of social capital (Putnam, 1993). See more: Nannestad, 2008. Some categories, however, are intertwined, for instance, emotional and moral perceptions of trust.

[125] Wolfe, 1976

[126] Frederiksen, 2011: 8

[127] Markova, 2004: 3-4; Habermas, 1984

social order and lead to social and economic integration ("*solidarity*"). Durkheim believed that trust is a ground on which collective conscience is built, and these two concepts go hand in hand ensuring social order by putting moral constraint on individuals' actions[128]. Trust facilitates collective behaviour and actions, as it organises our choices according to certain habits and cultural norms we are used to and do not need to reflect upon all the time. The moral approach to trust has been dominant in the so-called *Weberian* sociology, where trust is perceived as an inherent, religion-based feature of culture. Indeed, Simmel, following a Weberian way of thinking, was the first scholar who integrated and analytically conceptualised trust as a sociological subject in his two main studies: Philosophie des Geldes[129] and Soziologie[130]. According to Simmel, trust is evidently one of the most important synthetic forces within society and without it society itself would disintegrate[131]. Metaphorically speaking, trust helps to convert the Hobbesian state of nature from something that is brutish and selfish into something that is more efficient, pleasant and altogether more peaceful[132]. To sum up, seen from a moral point of view, trust holds an element of "*faith*" which is endogenous, attached to religion and even mystical, having nothing to do with practical knowledge or personal experiences[133]. Trust has a unique, metaphysic nature, therefore some theo-

---

[128] Misztal, 1996: 46

[129] 1900, The Philosophy of Money

[130] 1908, Sociology: Investigations on the Forms of Sociation

[131] 1950: 326

[132] Newton, 2001: 202

[133] Simmel, 1950:318; Uslaner, 2002

rists expect that trust is relatively consistent over different phases of individual's life[134].

Alongside the moral and rational interpretations of trust, Sztompka[135] distinguishes the third perception of trust as a cultural rule. From this point of view, trust is conceived as a prescribed norm: individuals are expected to behave according to the norm of trust/distrust. Put it differently, an individual's decision to trust or distrust someone depends on the pre-existent cultural context, where normative rules encourage or refrain from trust. Trust is not treated as an entirely voluntary individual choice (based on morality or rational calculations), but as a cultural phenomena, or social fact, in the sense of Durkheim. Sztompka[136] points out that if rules demanding trust are shared by a community and trust is a perceived external obligation, then these rules exert a strong pressure on actually giving trust. And the other way around, if a culture implies suspicion and distrust, the rules of this culture impose the withdrawal of trust. In other words, people take into account a general level of trustworthiness in a particular context and based on that they make their decision to trust others or not[137]. Trust as a cultural rule might transform the origins of trust, for instance, decrease rational calculations, and vice versa, suspend the moral inclination of trust.

However, generalised trust can actually exist when it is founded on a moral, rather than rational, calculative basis. Seen like this,

---

[134] e.g. Seligman, 1997; Uslaner, 2002

[135] 1999: 66–68

[136] 1999: 66

[137] Hardin, 2002

generalised trust appeals to a regularly honest behaviour of a trustee.

But, *what are determinants of social trust?*

First, at the individual level, social trust is more often a feature of people who are educated and have a higher socio-economic status[138]. Trust increases with education because more educated people presumably are more knowledgeable, tolerant and less bounded by prejudices.[139] Offe[140] notices that the rich, the more powerful and well-informed people are more inclined to trust, because they are less vulnerable to the potential losses of trust. These people are more likely to survive the disappointment of falsified trust investment, because they can switch to alternative resources (like money or power), whereas people lacking these resources are in a greater risk to suffer from a breakdown of the trust relations. Simply put: poverty makes people risk-averse. Similarly, Delhey and Newton[141] note that: "*Those who have been treated kindly and generously by life are more likely to trust than those who suffer from poverty, unemployment, discrimination, exploitation and social exclusion*". Furthermore, the better-off are more trusting because they are surrounded by trustworthy people like themselves[142]. This aspect is very important, as the question that measures social trust ("generally speaking, would you say that most people can be trusted or that you can't be too careful in dealing with people") is about how individuals judge the trustworthiness of

---

[138] Inglehart, 1997; Uslaner, 1999; Putnam, 2000

[139] Vasilache, 2010

[140] 1999

[141] 2003: 96)

[142] Zmerli and Newton, 2011

others[143]. Hence, contextual conditions are no less important than individual propensities. Second, at the societal level, the literature suggests that generalised social trust is a characteristic of individualist cultures, which stand in opposition to collectivist societies[144]. While collectivist cultures are in-group and family centred, individualistic values promote equality across different sections of society and thus encourage a generalised sense of trust. These attitudes are especially common to Protestant cultures[145]. Inglehart[146] argues that modern, economically developed societies steadily turn from materialist or security based values, to post-materialist values such as individual's autonomy and self-expression.

Due to rising economic prosperity individuals put survival values in the second plan and give priority to individual-improvement values.

Social trust with its modern connotation strengthens the post-materialist normative set (and vice versa), which is oriented towards the society based on humanist ideas. Social trust evolved along such emphasised values as personal freedom, a sense of control and efficacy, promoting a healthy environment, human rights, peace, equality, life satisfaction, and happiness. At first glance, these world views do not seem to be directly related to trust, but trust is in fact the essential component of social prosperity, leading to common understanding and solutions for "making the world a better place".

---

[143] Zmerli and Newton, 2011:1

[144] Triandis, 2004

[145] Wolfe, 1989; Inglehart, 1997

[146] 1997; 1999; 2008

As Monroe[147] puts it, social trust contributes to creating a global identity that emphasises inclusive human values.

Another important societal prerequisite that might encourage or reduce social trust is social/ethnic diversity. Some authors[148] claim that diversity increases the odds of having social contacts with different people, so people become accustomed to "strangers". But the adverse effect is also possible: diversity (especially, conflict based) might deter people from cooperation and shelter them under their own circles[149]. Social capital theorists[150] claim that social trust is generated by the interaction of individuals in civic networks.

We learn to trust when communicating with others in civic associations that represent a segment of a society.

The contacts with the members of an association allow for the process of generalisation of social trust in people you do not know[151]. It also teaches us how to safely project trust in others and it has a heuristic function: *"if others have learned similar lessons, then trust will in fact become generalised throughout the society"*[152].

However, this argument has received criticism as well[153]: firstly, not all civic associations breed social trust, and secondly social trust might even decrease because of the associational "localism" and distance created towards others outside the networks. Very intere-

---

[147] 1991

[148] Allport, 1954; Stolle, Soroka, and Johnston, 2008

[149] Delhey, Newton, and Welzel, 2011; Hooghe et al., 2009

[150] Putnam, 1993, 2000; Brehm and Rahn, 1997; Norris, 2002a

[151] Stolle, 2002

[152] Levi, 1996: 48

[153] Newton, 2001; Uslaner, 2002; Stolle, 1998

sting evidence in this regard was demonstrated by Stolle (1998), who found out in her research that joining associations indeed increases social trust for a short time, but in a long run social trust tends to shrink between the members of different civic associations. She concludes that in the initial stage members strengthen trust bonds with each other, but after some time their trust does not generalise to "outsiders" of a particular association. Uslaner's idea[154] is that activities do not lead to trust, but vice versa, trust leads to different activities. It means that if you are particularised truster, you are likely to choose cooperation in narrow groups with people who are similar to you. And if you are a generalised truster, you tend to participate in different types of associations, cutting across social cleavages. These associations might include people who are different from you, but you need to agree with them upon important decisions. Another group of authors[155] suggests that social trust stems from informal social networks, or, face-to-face interactions with people you know — family, friends, neighbours. These tight and strong relations, according to the authors, actually teach us "the virtue of trust".

Thirdly and finally, the quality of institutions and legal norms are thought to be a crucial prerequisite of social trust — or even the factor which makes trust "generalised" as such[156]. Institutions and fair legal systems are thought to reduce risks related to trust and create favourable conditions for spreading trust at the generalised level. First, efficiently protected rights of individuals and well-func-

---

[154] 2002

[155] Foley and Edwards, 1998; Delhey and Newton, 2005

[156] Coleman, 1990; Levi, 1996; Offe, 1999, Yamagishi and Yamagishi, 1994; Warren, 1999, Misztal, 2001; Rothstein and Stolle, 2008

tioning mechanisms of damage compensation make it more likely for individuals to trust others compared with the social contexts where rights are abused and legal norms are inefficient[157]. In this sense, a democratic environment creates an ability to anticipate the future and thus trust could become a "rational gamble", a voluntary choice. In totalitarian societies, contrarily, trust may cause huge losses because of the lack of anticipation. Second, as Offe[158] observes, institutions provide normative reference values that can be relied on in order to justify the rules created by institutions. Institutions are endowed with a certain spirit, a certain moral setting which prescribes preferred ways to conduct for people in the community. If institutions are available to effectively ensure the compliance of the citizens to the values, it implies that citizens should trust their fellows as they are involved in the same institutions. This involvement creates commitment to the norms and values represented by these institutions[159]. However, if citizens feel in doubt about the moral inclination of the institutions, they most probably will neither elaborate loyalty towards them, nor towards fellow citizens.

In academic writing, social trust and institutional trust are sometimes conflated within the more abstract notion of political trust[160]. However, these are two different analytical categories. Thus we separate these notions in this dissertation and refer to social trust as trust between people and political trust as trust in political institu-

---

[157] Sztompka, 1999; Warren, 1999; Jong-sung You, 2012

[158] 1999

[159] Offe, 1999: 65-67

[160] Heywood and Wood, 2011: 148

tions. Compared to social trust, political trust[161] rests on a vague and partial understanding[162], because the relationship between truster (individual) and trustee (institution, politician) is not direct or equal. Political trust can be learned through primary contact (personal experience with the institutions), but as Newton[163] observes, it is usually learned indirectly and at a distance — through media.

Indeed, media play a crucial role in establishing pillars of political trust, providing us with information to make judgments about politicians and institutions. We may not know the people in government personally, but we believe that we have sufficient information to make expectations and judgments about them. Unlike some forms of social trust, political trust is always related to expectations, so in this sense it is a strategic, but not moralistic trust (Uslaner, 2002). Conceptualised in this manner, political trust applies to actions of some institution or politician being in line with my normative expectations, even if we do not check these actions permanently. Therefore, political trust is sometimes understood as a psychological orientation, having both affective and evaluative aspects[164]. There is a rich literature discussing the dialectics of political trust and liberal democracy, and we would like to make a brief detour into that to argue how social and political trust are related (we find this philosophical literature helpful in this regard). To begin with, in

---

[161] Some authors think that, when we talk about institutions and systems, the term trust is inappropriate and should be replaced with more neutral terms such as reliance or confidence (for instance, Dalton, 1999; Sapsford and Abbott, 2006). On the other hand, the term political trust is also widely used and discussed (Mishler and Rose 2001, Newton, 2007; Van der Meer, 2010).

[162] Giddens, 1990: 179

[163] 2001: 205

[164] Norris, 1999

liberal political thinking, trust is a fairly controversial notion[165]. Although sociological theories consider political trust and democracy as mutually supportive, according to liberal philosophy, the roots of the liberal system lie, in fact, in distrust. The French philosopher Pierre Rosanvallon[166] maintains that distrust is a natural and legitimate component of democracy, and it functions as a protective mechanism, enabling society to control the democratic processes alongside the formal rules. Following the thought of liberal philosophy, distrust is a necessary condition for institutions not to override their authority or abuse the rights and freedoms of ordinary citizens. Historically, the institutionalisation of distrust in the political system was tightly related to economic liberalism. The Constitution of the United States (1787) has primarily anchored the legal mechanisms of distrust in the realm of economics: it inscribes protective mechanisms on behalf of economic liberties against the intervention of the state in economic relations. These mechanisms have been transferred to the more abstract sphere of politics, first of all, by means of the "division of powers", allowing institutions to compete with each other for power and thus restraining each other's possibilities for systemic usurpation. Moreover, distrust is also institutionalised through additional "safeguards": a multi-party system, election rules, the right to competition, monitoring, and formalities that regulate the time span and periodicity of office terms[167]. However, the constitutional rules and formal safeguard mechanisms alone are not sufficient to avoid the abuse of power by institutions.

---

[165] Warren, 1999; Hardin, 2006; Rosanvallon, 2008

[166] 2008

[167] Benn and Peters, 1959: 281

Democracy also encompasses a wide range of resistance forms that could be used by society against the government, and these forms surpass the limits of formal rules[168].

People are not only just voters (passive participants), but also active quality controllers (critical citizens) of the political system.

The philosopher refers to this kind of interaction, when citizens maintain the control levers through protest, as "counter democracy"[169]. Permanent distrust in the political system exerted by the people is arguably one of the fundamental substantial ideas for democracy to truly work. Braithwaite and Levi[170] label the democratic project as "institutionalised distrust". Hardin[171] acknowledges that distrust is one of the key conditions for modern democracy. Power inequality between state institutions and common people is too immoderate (unbalanced), but we have no alternatives to these institutions. We are dependent on them. Institutionalised distrust implicates institutional accountancy — so called "agencies of accountability" that may enforce trustworthiness of the system. These agencies (courts, police, controllers, examination boards, media, etc.) put pressure on persons, institutions, or systems that are our targets of trust[172]. Yet enforcement agencies must be trustworthy themselves. If citizens do not trust these agencies, they will not trust their officials to fulfil their duties[173]. It should be clarified that institutionali-

---

[168] Warren 1999; Rosanvallon, 2008

[169] Rosanvallon, 2008)

[170] 1998

[171] 2006: 152

[172] Sztompka, 1999: 47

[173] Dasgupta, 1988: 50

sed political distrust (substantial distrust) is not the same as political distrust in concrete political institutions (formal distrust). To avoid confusion, it is expedient to differentiate between these two-forms of political distrust.

Formal political (dis)trust would be expressed towards, for instance, the parliament, the government, or concrete politicians.

Substantial political distrust refers to the permanent distrust (or "healthy" suspicion) of institutional politics or of the system as such.

Hence, in liberal thinking, political participation — voting, writing petitions, demonstrations, and boycotts — is the expression of substantial political distrust. We participate in elections in order to control the powers of institutions and express our substantial distrust in them. Thus, exposing substantial political distrust does not imply that we also feel formal political distrust at the same time.

Following this distinction, many authors actually believe that social trust and formal political trust are mutually reinforcing[174]. Some of them[175] even think that political trust indeed gives an impulse for social trust to emerge. It is argued that trust in a certain system as a set of values empowers us to trust citizens of this system, as we all belong to the same setting of normative rules and general morality. Farrell and Knight[176] suggest that institutions create rules and sanctions for people to behave in a trustworthy manner, thereby fostering trust. Similarly, Levi[177] argues, "Governments pro-

---

[174] Gambetta, 1988; Burt, 1992; Putnam, 1993, Sztompka, 1999; Misztal, 1996; Levi, 1998

[175] Sztompka 1999, Warren, 1999; Rothstein and Stolle, 2002

[176] 2003

[177] 1996: 51)

vide more than the backdrop for facilitating trust among citizens; governments also influence civic behaviour to the extent they elicit trust or distrust towards themselves". We can also talk about the positive effect of social trust on political trust. As Putnam[178] observes, if people are willing to trust strangers, they will also trust politicians and political institutions. However, there is empirical evidence questioning the link between social and political trust. For instance, institutional theories argue that social trust has nothing to do with political trust and the latter depends on citizens' evaluations of the political and economic performance of the government[179].

In other words, political trust is a consequence of institutional performance, but not a result of social trust.

This insight is also displayed by empirical research showing that political and social trust are weakly correlated[180]. For instance, referring to extensive research, Uslaner[181] points out that social trust has no significant influence on political trust at the individual level. Moreover, according to this research, particularised trust even has a negative effect on trust in government[182]. It would confirm Fukuyama's claim that private (or particularised) trust is a substitute for the lack of institutional trust.

To sum up, political trust and social trust are different in their foundations and functions, and there is no clear empirical evidence whether these two types of trust are directly connected. On the

---

[178] 2000

[179] Parry, 1976; Mishler and Rose, 2005; Zmerli and Newton, 2008

[180] Newton, 1999a; Kaase, 1999; Uslaner, 2002; Delhey and Newton, 2003

[181] 2002

[182] The same results were found in the study of Zmerli and Newton, 2011

other hand, social trust and political trust relate to the same phenomenon — democratisation. Both types of trust emerged due to the institutional and cultural shifts that took place along the modernisation processes. In a more systemic way, social trust and political trust accomplish and sustain each other. These forms of trust are both positively associated with life satisfaction and happiness, education, income and civic engagement[183]. Similarly, although emphasising the different nature of political trust, Newton and Norris[184] argue that this relationship is apparent at the aggregated societal level. Thus, it means, that at the contextual level, in the long run social and political trust are likely to adjust to each other. Newton[185] refers to Finland as an example: in some historical moment, this country suffered from low levels of political trust, but it was soon recovered with a strong help of social trust. At the same time Newton[186] hypothesises, that a country with equally severe political distrust and low social trust is likely to experience a greater problems in building or recovering political trust.

So far, we have discussed a series of theoretical aspects about political participation, efficacy, social trust and its associate — political trust. Now we aim at integrating these variables in a coherent framework and at demonstrating the relationships between them.

Social capital

The social capital literature most particularly addresses the reciprocal relations between social trust and collective actions. Broadly

---

[183] Zmerli and Newton, 2011:77

[184] 2000

[185] 2001

[186] 2001: 210

speaking, this relationship constitutes the core of the social capital concept, which aims to explain how social interactions may strengthen democratic institutions, or "make democracy work". In his ground-breaking study *Making Democracy Work*, Putnam defines social capital (the term was coined by Bourdieu[187]) as a possession which does not contain material assets, but connections among individuals: it is *"features of social organization such as trust, norms, and networks that can improve the efficiency of the society by facilitating coordinated actions"*[188]. As it clearly flows from the definition, trust is an integral component of social capital[189]. Trust works in concert with norms/ obligations and social networks: "the causal arrows among civic involvement, reciprocity, honesty and social trust are as tangled as well-tossed spaghetti". This relationship is circular: the more an individual trusts, the more s/he tends to cooperate and is exposed by civic norms, and vice versa, civic engagement and shared norms lead to new trust relationships. It constitutes a so-called social spiral with social trust and political action at both ends. Hence, in Put-

---

[187] Bourdieu (1986: 241) defines three forms of capital: "Depending on the field in which it functions, and at the cost of the more or less expensive transformations which are the precondition for its efficacy in the field in question, capital can present itself in three fundamental guises: as economic capital, which is immediately and directly convertible into money and may be institutionalised in the forms of property rights; as cultural capital, which is convertible, on certain conditions, into economic capital and may be institutionalised in the forms of educational qualifications; and as social capital, made up of social obligations ('connections'), which is convertible, in certain conditions, into economic capital and may be institutionalised in the forms of a title of nobility".

[188] 1993: 167

[189] There is a disagreement on the definition of social trust and some authors claim that Putnam's proposed conceptualisation is too narrow, favouring only social networks. Halpern (2005: 9-19), for instance, systemises the existing definitions of social capital into a theoretical triangle which includes social networks, norms (in a wider sense, reciprocity, trustworthiness and trust) and sanctions (which also might include institutions). Seen from this angle, trust "loses" its role as separate component of social capital and is conceived as an integral part of norms. Different dimensions of social capital, however, are intertwined (Hardin, 2002)

nam's view trust is one of the mechanisms that produces and maintains social capital.

In line with the concept, active membership in organizations and involvement in voluntary associations are considered to be crucially important to transform trust from an individual characteristic to a collective resource.

Indeed, emphasis here is put on voluntary organizations (with an institutional setting), but not private circles like friends or family. Formal social networks are suggested to function as "schools of democracy" where "citizens learn to participate by participating". The argument of Almond and Verba[190] states *"Individuals can be expected to generalise from experiences outside political life to politics; if they have participated within non-political authority structures they will expect to do so in the political sphere also"*.

Hence, when citizens participate in small-scale civic associations, they are taught habits of cooperation and thus are able to socialise into larger political involvement.

Social organizations provide resources necessary for collective actions. The horizontal networks ensure that these resources, both material and cognitive/psychological, could be equally accessed by citizens. Subsequently, the resources and the shared norms (as social trust) empower people to solve more complicated problems in dealing with the institutional structures. So here the connection between social trust and political efficacy enters. Putnam observes that in more trustful communities a sense of political efficacy of ordinary citizens is higher. He explains this phenomena referring to the egalitarian nature of the society. If the societies' structure is horizontal and based on trust, people feel more powerful and capable

---

[190] 1963

of influencing political issues, because they would expect that other people would behave similarly to them. This idea implies that political efficacy in the social capital concept indeed connects the ends of social spiral, that is, social trust and political action. In contrast, in vertical and distrustful societies people perceive themselves being exploited, submissive and dependent, therefore their sense of political efficacy is constrained.

Conclusion: the relationships between social trust, political efficacy, and participation

Social trust functions as a mobilising force of movement participation. As Coleman[191] states, "*a group whose members manifest trustworthiness and place extensive trust in one another will be able to accomplish much more than a comparable group lacking that trustworthiness and trust*". Thus, we assume that social trust gives one initiative to take part in a protest action. Implicitly the idea also suggests that political efficacy is an explaining mechanism between social trust and protesting.

The alternative approach to the connection between trust and participation posits that social trust may work as "double-edged sword", meaning that trusting people can remain passive, because they believe that others can be trusted to participate (and "do the job") for them[192].

The scholarly literature also distinguishes between different associations of social trust and the modes of political participation, for instance, empirically supporting the negative links between social trust and participation in political parties/campaigns[193]. However, drawing from the social capital literature, the initial assumption is

---

[191] 1990: 304

[192] (Pattie, Seyd, and Whiteley, 2003: 458)

[193] (Hooghe and Marien, 2013; Hooghe and Quintelier, 2014; Van der Meer and Van Ingen, 2009)

that the relationship between social trust and political participation, despite their modes, follows the same logic as with civic participation.

Finally, the mother of all questions!

To what extent is the effect of social trust on participation mediated by political efficacy?

The question is why it is important to focus our attention particularly on the mediation effects of internal and external political efficacy in the analysis of the relationship between social trust and political participation. This debate centres on social capital and political capital literatures, explaining how social trust, as a social attitude, can convert into political skills. So it is important from the perspective that individual's social environment matters for his/her political inclinations and decisions.

In other words, the mediation effect would demonstrate that social resources could be transformed into individual political resources.

As we have already discussed, the literature on how social trust exactly affects political participation is quite ambiguous. Some studies find a positive effect of social trust on most forms of political participation[194], while others come across relatively weak, non-existent or even negative relationships, depending on the forms of political activity[195]. Generally, voting and movement politics in most studies occur to be positively associated with social trust, while the relationship with other institutionalised forms of participation is less clear. Moreover, social capital theories also suggest that social trust

---

[194] Kaase, 1999; Putnam, 1993; 2000; Rossteutscher, 2008: 228; Marien and Christensen, 2013

[195] Millner, 2000; Van Deth, 2000; Muhlberger, 2003; Rubenson, 2005; Armingeon, 2007; Hooghe and Marien, 2013; Hooghe and Quintellier, 2014

might have an indirect effect on political participation mediated by political efficacy. For instance, in *Making Democracy Work*, Putnam notices that social trust empowers people to solve more complicated problems in dealing with the institutional structures. So, here the pivotal connection between social trust and political efficacy enters.

In the civil society or civic culture literature, social trust, efficacy and participation are interlinked less explicitly. Here, social trust is prescribed to a certain set of mutual obligations together with political participation, respect for human rights, citizenry duties, perceptions of the common good and common responsibility. In *Civic Culture*, for instance, Almond and Verba[196] argued that subjective competence, participation, self-empowerment and social trust are connected through the socio-economic status and education level, which promote political efficacy, a sense of social trust and political activity. But as it was already mentioned, the research on this relationship is quite scarce, except for a few studies[197], and some other theoretical aspects[198]. Other authors provide some hints on how social trust positively contributes to political literacy and awareness[199], self-confidence, and also to political support and political trust[200].

Drawing on the literature, we consider three indirect positive effects of trust on participation. First, while placing trust in other individuals and receiving this trust reciprocally, we feel a moral duty

---

[196] 1963

[197] for instance, Hsung, 2014; Anderson, 2010; Van Deth and Scarbrough 1998

[198] Van der Meer and Van Ingen, 2009), Hooghe and Marien (2013)

[199] Milner, 2002

[200] Newton, 1999a; Norris, 2002a

to be involved in common affairs[201]. It is argued that social trust leads individuals to participate in the pursuit of the commons, as trust provides us with assurance that political action will be appreciated and at least potentially effective. In return, successful cooperation based on trust gives people satisfaction in what they do[202]. Social trust endows us with the feeling of being part of a larger community and provides us with a sense that we, people, can make a difference. This first point refers to both external and internal efficacy. Second, while trusting, people more likely acquire political knowledge and information in general about the subjects to be acted on and the methods to employ such actions (internal efficacy). Third, trust empowers us politically as we believe that institutions we deal with are fair and people we trust are going to behave by the rules.

When we trust people, we expect them to act in a similar manner and it gives us a sense of control (predictability).

Sustained confidence in the motives of others encourages and facilitates participation[203].

If we are sure that the rules are not going to be violated, we feel more certain about involvement in political life (external efficacy).

Unfortunately, as soon as people take on the role of politician and decision maker, they change opinion and forget what causes they were elected for.

So, we should reinforce also our capacity to choose our delegated trustees.

---

[201] Putnam, 1993; 2000; Seligman, 1997

[202] Putnam, 1993

[203] Kwak, Shah, and Holbert, 2004

INDEX

Acknowledgments
And
Heartfelt thank to

Brian Czech, has a B.S. in wildlife ecology from the University of Wisconsin-Madison, an M.S. in wildlife science from the University of Washington, and a Ph.D. in renewable natural resources from the University of Arizona.

David Batker, directs the APEX Center for Applied Ecological Economics.

Dr. Herman E. Daly is currently Professor at the University of Maryland, School of Public Affairs.

Dr. Joshua Farley has Ph.D. in Agricultural, Resource and Managerial Economics from Cornell University in 1999

To all resources, researchers, scholars, enthusiasts, professionals, volunteers mentioned in the work

Many heartfelt thank you, to everyone!

Selected Readings and Links

General Readings

French, Hilary. Vanishing Borders. New York: W. W. Norton, 2000.

Paehlke, Robert C. *Democracy's Dilemma: Environment, Social Equity and the Global Economy.* Cambridge, MA: MIT Press, 2003.

Visions for a Sustainable Future

Albert, M., Cagan, L., Chomsky, N., Hahnel, R., King, M., Sargent, L., and Sklar, H. *Liberating Theory*, South End Press, Boston, 1986

Albert, M. and Hahnel, R. *Looking Forward: Participatory Economics for the Twenty First Century*, South End Press, Boston, 1991

Ekins, Paul. *"The Sustainable Consumer Society: A Contradiction in Terms?"* International Environmental Affairs 3.4 (1991): 243-258.

Goodland, Robert, H. Daly and S. El Serafy (eds.). *Population, Technology and Lifestyle: The Transition to Sustainability.* Washington: Island Press, 1992.

Odum, Howard and E. Odum. *A Prosperous Way Down.* Boulder, CO: University Press of Colorado, 2001.

Schor, Juliet and B. Taylor. *Sustainable Planet: Solutions for the Twenty-First Century.* Boston: Beacon Press, 2002.

Speth, James. *Red Sky at Morning.* New Haven: The Yale University Press, 2004.

Williamson, T. *What Comes Next?* National Center for Economic and Security Alternatives, Washington, D.C., 1997

Schneider Stephen, *"Laboratory Earth"* - New York: Basic Books, 1997

www.steadystate.org

Daly, Herman and J. Cobb. *For the Common Good: Redirecting the Economy toward Community, the Environment and a Sustainable Future.* Boston: The Beacon Press, 1989

Daly, Herman and J. Farley. *Ecological Economics: Principles and Applications.* Washington: Island Press, 2004.

Roodman, David. *The Natural Wealth of Nations.* New York: W. W. Norton & Company, 1998.

Well-Being and Happiness

Diener, E. and C. Diener. *"The Wealth of Nations revisited: Income and the quality of life,"* Social Indicators Research 36 (1995): 275-286.

Diener, E. and R. E. Lucas. *"Personality and Subjective Well-being."* In D. Kahneman, E. Diener and N. Schwartz, (eds.). Understanding Well-being: Scientific Perspectives on Enjoyment and Suffering. New York: Russell Sage, 1998.

Kasser, Tim. *The High Price of Materialism.* Cambridge, MA: The MIT Press, 2002.

Lane, R.E.. *The Loss of Happiness in Market Democracies,* Yale University Press, New Haven, 2000

Smil,V. *Energy at the Crossroads,* MIT Press, Boston, 2004: pp. 97-105

Hanaa Abdelaty Hasan Esmail & Nedra Nouredeen Jomaa Shili - *The Relationship between Happiness and Economic Development in KSA: Study of Jazan Region*

Public Policies

Anielski, Mark. *"Fertile Obfuscation: Making money whilst eroding living capital."* Presented at 34th Annual Conference of the Canadian Economics Association, June 2-4, 2000: Vancouver, http://www.pembina.org/pdf/publications/fertile.pdf.

Cohen, Joel. *How Many People Can the Earth Support?* New York: W. W. Norton, 1995.

Daly, Herman and J. Farley. *Ecological Economics: Principles and Applications.* Washington: Island Press, 2004.

Daly, Herman and J. Cobb. *For the Common Good: Redirecting the Economy toward Community, the Environment and a Sustainable Future.* Boston: The Beacon Press, 1989.

Daly, Herman. *Beyond Growth: The Economics of Sustainable Growth.* Boston: Beacon Press, 1996.

Daly, Herman. *Steady-State Economics (2e).* Washington: Island Press, 1991.

Daly, Herman and K. Townsend (eds.). *Valuing the Earth: Economics, Ecology, Ethics.* Cambridge, MA: MIT Press.

Daly, Herman (ed.). *Ecological Economics and the Ecology of Economics.* Northampton: Edward Elgar, 1999.

Davidson, E. You *Can't Eat GNP: Economics as If Ecology Mattered.* Cambridge, MA: Perseus, 2000.

French, Hilary. *"Investing in the Future: Harnessing Private Capital Flows for Environmentally Sustainable Development."* *Worldwatch Paper* (139), February 1998.

Macower, Joel and D. Fleischer. *Sustainable Consumption and Production Strategies for Accelerating Positive Change.* New York: Environmental Grantmakers Association, 2003.

Mastny, Lisa. *"Purchasing Power: Harnessing Institutional Procurement for People and the Planet."* Worldwatch Paper (166) July 2003.

Michalos, Alex. *Good Taxes.* Toronto: Dundurn Press, 1997.

National Research Council. *Nature's Numbers: Expanding the National Income Accounts to Include the Environment.* Washington: National Academy of Sciences, 1999.

Organization for Economic Co-operation and Development. *Environmentally Related Taxes in OECD Countries: Issues and Strategies.* Paris: OECD, 2001.

Roodman, David. *"Getting the Signals Right: Tax Reform to Protect the Environment and the Economy."* Worldwatch Paper (134) May 1997.

Roodman, David. *The Natural Wealth of Nations.* New York: W. W. Norton & Company, 1998.

Von Weizsacker, Ernst and J. Jesinghaus. *Ecological Tax Reform.* London: Zed books, 1992.

Biermann, Frank. *"The Case for a World Environment Organization"* - *Environment* 42.9 (2000): 23-31.

Charnovitz, Steve. *"A World Environment Organization"* - Columbia Journal of Environmental Law 27.2 (2002): 323-363.

Chertow, M. R. and D. C. Esty (eds.). *Thinking Ecologically: The Next Generation of Environmental Policy.* New Haven: Yale University Press, 1997.

Choucri, N. (ed.). *Global Environmental Accord: Strategies for Sustainability and Institutional Innovation.* Cambridge, MA: MIT Press, 1993.

Environmental Law Institute. *Harnessing Consumer Power.* Washington: ELI, 2003.

Esty, Daniel and M. Ivanova (eds.). *Global Environmental Governance: Options and Opportunities.* New Haven: Yale School of Forestry and Environmental Studies, 2002.

Esty, Daniel. *"Toward Data-Driven Environmentalism: The Environmental Sustainability Index,"* Environmental Law Reporter: News and Analysis 31.5 (2001): 10603.

Haas, Peter M., R. Keohane and M. Levy. *Institutions for the Earth: Sources of Effective International Environmental Protection.* Cambridge, MA: MIT Press, 1993.

Juma, Calestous. *"The Perils of Centralising Global Environmental Governance,"* Environment Matters: Annual Review, July 1999-June 2000, 13.

Weiss, E. and H. Jacobson. *Engaging Countries: Strengthening Compliance with International Environmental Accords*. Cambridge, MA: MIT Press, 1988.

Young, Oran (ed.). *The Effectiveness of International Environmental Regimes*. Cambridge, MA: MIT Press, 1999.

Economics for Community

Czech, Brian. Shoveling Fuel for a Runaway Train. Los Angeles: University of California Press, 2000.

Daly, Herman and J. Farley. *Ecological Economics: Principles and Applications*. Washington: Island Press, 2004.

Daly, Herman and J. Cobb. *For the Common Good: Redirecting the Economy toward Community, the Environment and a Sustainable Future*. Boston: The Beacon Press, 1989.

Daly, Herman. *Beyond Growth: The Economics of Sustainable Growth*. Boston: Beacon Press, 1996.

Daly, Herman. *Steady-State Economics (2e)*. Washington: Island Press, 1991.

Durning, Alan and Y. Bauman. *Tax Shift*. Seattle: Northwest Environment Watch, 1998.

Nadeau, Robert L. *The Wealth of Nature: How Mainstream Economics Has Failed the Environment*. New York: Columbia University Press, 2003.

Roodman, David. *"Getting the Signals Right: Tax Reform to Protect the Environment and the Economy."* Worldwatch Paper (134) May 1997.

Roodman, David. *The Natural Wealth of Nations*. New York: W. W. Norton & Company, 1998.

Sustainable Business Practices

Bakan, J. the Corporation: *The Pathological Pursuit of Profit and Power,* Viking Canada, Toronto, 2004

Elkington, John. *The Chrysalis Economy.* Oxford: Capstone Publishing Inc., 2001.

Hawken, Paul, A. Lovins and L. Lovins. *Natural Capitalism.* New York: Little, Brown and Co., 1999.

Hawken, Paul. *The Ecology of Commerce.* New York: HarperCollins, 1993.

Holliday, Chad and J. Pepper. *Sustainability Throughout the Market: Seven Keys to Success.* Geneva: WBCSD, 2001.

International Institute for Environment and Development and WBSCD. *Breaking New Ground: Mining, Minerals and Sustainable Development.* Geneva: WBCSD, 2002.

Korten, David. *When Corporations Rule the World.* San Francisco: Barret-Koehler Publishers, 1996.

McDonough, William and M. Braungart. *Cradle to Cradle: Remaking the Way We Make Things.* New York: Farrar, Straus and Giroux, 2002.

Von Weizsacker, Ernst, A. Lovins and L. Lovins. *Factor Four: Doubling Wealth*, Halving Resource Use. Earthscan Publications Ltd., 1999.

Willard, Bob. *The Sustainability Advantage.* Gabriola Island, BC: New Society Publishers, 2002

World Resources Institute, UN Environment Programme and WBCSD. *Tomorrow's Markets: Global Trends and Their Implications for Business.* Washington: World Resources Institute, 2002

Lifestyle Solutions

Dominguez, J. and V. Robin. *Your Money or Your Life*. US: Penguin Books, 1992.

Elgin, D. Voluntary Simplicity. New York: William Morrow, 1993.

Halwell, Brian. *"Home Grown: The Case for Local Food in a Global Market."* *Worldwatch Paper* (163) November 2002.

Lewis Akenji (IGES), Huizhen Chen - *A framework for shaping sustainable lifestyles - determinants and strategies.* United Nations Environment Programme, 2016

www.envisioninglifestyles.org

https://www.oneplanetnetwork.org

www.oneearthweb.org

The Good Life Goals by Futerra Sustainability Communications Ltd and 10-Year Framework of Programmes on Sustainable Lifestyles and Education Programme is licensed under CC BY-ND 4.0

Trusting people

Teodora Gaidytė, 2015 - Vrije Universiteit - *Explaining political participation in mature and post-communist democracies: Why social trust matters?*